Diagnostic Liquid-Based Cytology

Rana S. Hoda • Christopher VandenBussche
Syed A. Hoda

Diagnostic Liquid-Based Cytology

Springer

Rana S. Hoda, FIAC
CBL Path
Rye Brook, NY
USA

Christopher VandenBussche
Department of Pathology
Johns Hopkins University Department of Pathology
Baltimore, MD
USA

Syed A. Hoda
New York Presbyterian Hospital
Weill Cornell Medical College
New York, NY
USA

ISBN 978-3-662-53903-3 ISBN 978-3-662-53905-7 (eBook)
DOI 10.1007/978-3-662-53905-7

Library of Congress Control Number: 2016963680

Printed on acid-free paper

This Springer imprint is published by Springer Nature
The registered company is Springer-Verlag GmbH Germany
The registered company address is: Heidelberger Platz 3, 14197 Berlin, Germany

Drs. Rana and Syed Hoda dedicate the book to Sehyr and Raza, the light of their lives. Dr. Christopher VandenBussche dedicates it to his beloved wife, Cherry.

Preface

Liquid-based cytology preparations are currently the standard of care for gynecological cytology and are being increasingly used for non-gynecological cytology. *Diagnostic Liquid-Based Cytology* serves as a handy guide to diagnostic cytopathology on liquid-based preparations.

Diagnostic Liquid-Based Cytology is lean enough to be easily read cover to cover within a reasonably short period of time. It is hoped that the reader will refer to this handbook often, or as the need arises, to learn more about liquid-based cytopathology preparations. The book is intended to readily help the reader with the interpretation on such material. Lastly, it is hoped that the readers will find helpful information throughout these pages for taking various proficiency and licensing examinations.

Rana S. Hoda
Christopher VandenBussche
Syed A. Hoda

Acknowledgments

The inspiration imparted to the authors by their many beloved teachers—including, in alphabetical order, Dr. Prabodh Gupta, Dr. Stephen Hajdu, Dr. Leopold G. Koss, Dr. Paul Peter Rosen, Dr. Richard J. Reed, and Dr. Juan Rosai—is gratefully acknowledged.

Ms. Patricia Kuharic provided masterly expertise in the preparation of the illustrated content of this book.

Contents

Authors

1. **Rana S. Hoda, M.D., F.I.A.C.**
   Director of Cytopathology
   CBL Path, 760 Westchester Avenue, Rye Brook, NY 10573
   T 914.698.5706, Email: Rhoda@cblpath.com

2. **Christopher VandenBussche, M.D., Ph.D.**
   Assistant Professor of Pathology and Associate Director of Cytopathology Division
   The Johns Hopkins Hospital, 600 N. Wolfe Street, Sheikh Zayed Tower, Baltimore, MD 21287
   T 410.955.1180, Email: cjvand@jhmi.edu

3. **Syed A. Hoda, M.D.**
   Professor of Clinical Pathology and Laboratory Medicine and Attending Pathologist of Weill Cornell Medical College-New York Presbyterian Hospital
   525 East 68th Street, Starr 10, New York, NY 10065
   T 212.746.2708, Email: sahoda@med.cornell.edu

Other Contributors

1. **William C. Faquin, M.D., Ph.D.: Chapter 8**
   Professor of Pathology, Harvard Medical School
   Director, Head and Neck Pathology
   Department of Pathology, Warren 219
   Massachusetts General Hospital
   55 Fruit Street, Boston, MA 02114
   T 617-573-3957, E-Mail: wfaquin@mgh.harvard.edu

2. **Mine Onenerk, MD: Chapter 8**
   Research Fellow, Head and Neck Pathology
   Department of Pathology, Warren 219
   Massachusetts General Hospital
   55 Fruit Street. Boston, MA 02114
   T 617-573-3957, E-mail: aonenerk@mgh.harvard.edu

3. **Scott A. Ely, MD: Chapter 10**
   Associate Professor of Pathology
   Department of Pathology
   Weill Cornell Medical College
   New York Presbyterian Hospital
   525 East 68th Street, Starr 715
   New York, NY 10065
   T 212-746-2442, Email: S12564@pol.net

Liquid-Based Specimen Collection, Preparation, and Morphology

1

Introduction

Liquid-based preparations (LBP) are increasingly being used for both gynecologic and non-gynecologic (non-gyn) cytology including fine needle aspiration (FNA). Although the diagnostic criteria used for conventional preparations can also be applied to LBP, an accurate interpretation of liquid-based slides requires familiarity with the background, architectural, and cytological alterations to avoid misdiagnosis.

The types of LBP include SurePath [(SP), formally known as AutoCyte PREP, BD Diagnostics TriPath, Burlington, NC, USA] and ThinPrep [(TP), Hologic, Marlborough, MA, USA] and Millipore Filter (Bedford, MA, USA) and Liqui-PREP (LGM International, Fort Lauderdale, FL, USA). Currently, only SP and TP have FDA approval since they are the more commonly used LBP. The TP and SP preparations were approved for cervicovaginal (Pap test) cytology in 1996 and 1999, respectively, and both have since also been used for non-gynecological cytology including fine needle aspiration (FNA). Both TP and SP are also FDA-approved for high-risk human papillomavirus (HPV) testing of Pap tests. HPV test can also be performed on SP specimens after laboratories have performed internal validation. The published results have been shown comparable HPV performance of SP to TP. All illustrations depicted in this book are either SP or TP. Details of these commonly used LBP have been extensively published in the literature and also appear in the subsequent section of this chapter and in organ-specific chapters.

In conventional smears, the collected patient sample is smeared on glass slides, and the collection device, containing some of the diagnostic cell sample, is discarded. Moreover, the collected sample is nonuniformly distributed on the glass slide and may span the entire surface area of the slide. In LBP, unlike the conventional smears, the acquired patient sample, from various anatomic sites, is rinsed in specially designated collection preservative media ensuring collection of potentially

R.S. Hoda et al., *Diagnostic Liquid-Based Cytology*,
DOI 10.1007/978-3-662-53905-7_1, © Springer-Verlag GmbH Germany 2017

100 % of the cell sample. The collection media for TP is CytoLyt (Hologic, Marlborough, MA, USA) and that for SP is CytoRich Red (BD Diagnostics TriPath, Burlington, NC, USA).

There are several advantages to LBP compared to conventional smears. Both LBP techniques reduce debris and cell clumps and also homogenize the specimen, increase specimen adequacy, result in uniform cell distribution in a smaller screening area (the screening area of a conventional smear is 2.5 × 1.5 cm, the size of the glass slide), show cleaner background, reduce slide screening time, remove air-drying artifact due to immediate liquid fixation, enhance cellular and nuclear details for improved diagnostic accuracy, and increase productivity. Moreover, residual material from LBP can be successfully used to process cell blocks, which provides additional diagnostic information, including architectural pattern. Both LBP and cell block material can also be utilized for ancillary studies such as immunocytochemistry, special stains, and molecular tests. If LBP are used for ancillary studies, additional slides can be prepared from the specimen vial. Advantages and disadvantages of various cytological preparations, technical details for the two LBP, and their general and specific cytological characteristics are provided in Figs. 1.1 and 1.2 and Tables 1.1, 1.2, 1.3, and 1.4. Step-by-step preparatory techniques for TP and SP also appear in this chapter.

Pathologists need to become familiar with LBP morphology due to alterations in background, architectural, and cytological features to avoid misdiagnosis. These differences in LBP, compared with conventional smears, are induced by the collection, fixation, and processing methods of LBP. Briefly, these cellular alterations compared with conventional smears include cleaner background with lack of or reduced background elements such as necrosis, blood, inflammation, mucin, and colloid, fragmentation of large clusters and papillary structures, artifactual aggregation of lymphoid cells, smaller cell size, and prominent nucleoli even in benign/reactive conditions. These cellular differences are described in detail below and with organ-specific chapters.

Liquid-Based Processing Techniques

SurePath (SP)

Principle

SP is a density gradient-based cell enrichment process. It is a semiautomated technique. The specimen is processed on the PrepStain processor. Multiple preparations can be made from a single specimen vial.

Specimen Preservation

Non-gynecologic specimens are preserved in CytoRich® Red Preservative Fluid (an ethanol-based medium which also lyses blood).

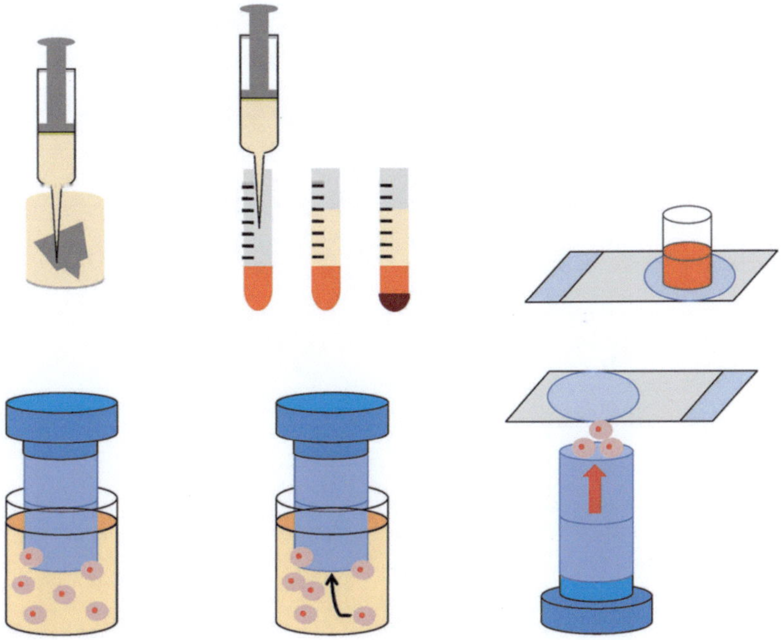

Fig. 1.1 SurePath™ and ThinPrep™ processing techniques
SurePath™ Technique: The sample is collected in CytoRich®, an ethanol-based proprietary preservative fluid which is vortexed prior to preparation. A syringe is used to disaggregate larger cell fragments. The specimen is then dispersed onto a density gradient reagent, a polysaccharide solution that acts to trap small particulates and debris. The specimen is centrifuged and then the cell pellet is resuspended. The PrepStain™ slide processor transfers the fluid to a settling chamber that rests on a positively charged glass slide. The PrepStain™ processor also automatically stains the slides.
ThinPrep™ Technique: A cylinder with a polycarbonate filter attached to one end is introduced into the specimen vial and gently rotated creating a current that disaggregates mucus, blood, and other debris, breaks up large cell clusters, and mixes and homogenizes the cell suspension. A gentle vacuum is then applied to the cylinder; most of the broken erythrocytes and debris pass through the filter pores, while the cells of interest adhere to the filter. The instrument monitors cell density across the filter and the flow rate decreases when cells are evenly distributed on the filter with minimal overlap. The cylinder then moves out of the specimen and is lightly pressed against a positively charged slide. A slight positive air pressure is applied to transfer the cells to the slide. The slide is immediately dropped into 95 % ethanol fixative. The slide is removed from the processor and may be stained either manually or by an automatic stainer (Modified from: Cibas ES, Ducatman BS. Liquid-based preparation methods. In Cytology. diagnostic principles and clinical correlations. 3rd ed. Philadelphia: Saunders; 2009. p. 5)

Slides

Pre-coated slides are provided by the company and are marked with a 13-mm diameter circle. The slides can also be freshly prepared in the laboratory for use within 48 h. The slides are coated with a modified poly-L-lysine and air-dried. These positively charged slides allow diagnostic cells to settle out of solution and adhere to the surface.

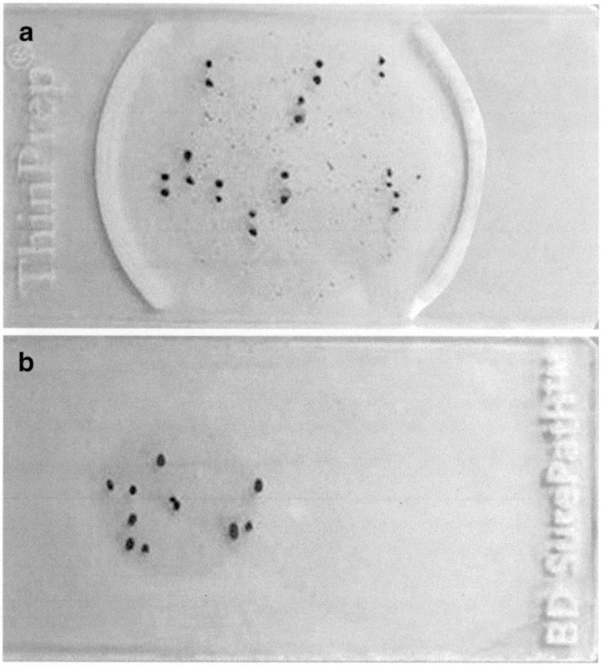

Fig. 1.2 (**a**) ThinPrep slide containing a circle 20 mm in diameter within which the cytological material is deposited. The specimen is prepared from a collection (preservation) medium which is methanol-based. (**b**) The SurePath slide contains a smaller circle, 13 mm in diameter, and is prepared from a collection medium which contains a mixture of methanol and ethanol. Because the specimen is consistently deposited within the well-defined diameters for each preparation, screening is more standardized and screening time is reduced

Processing

1. Pour one or two aliquots into 50 mL centrifuge tubes, cap securely, and centrifuge for 10 min at 600*g* to concentrate the specimen.
2. Decant the supernatant.
3. Vortex the cell button to form a homogeneous cell slurry. Transfer 10 mL of CytoRich® Red Preservative Fluid to the tube.
4. Specimen collection vials are vortexed and allowed to sit for at least 15 min. Vortex the specimen after sitting and transfer to a 12 mL tube. Please note that the sample may also be received in a vial with CytoRich collection fluid directly from the submitting clinician.
5. Centrifuge for 10 min at 600*g*.
6. Decant the supernatant with one firm motion, leaving residual supernatant with the specimen.
7. Vortex to homogenize the sample.
8. Load the centrifuge racks onto the PrepStain processor and run the program for non-gynecologic (non-gyn) specimens. A robotic arm transfers the fluid to a settling chamber which sits atop a modified poly-L-lysine-coated glass slide. The

Table 1.1 Principal advantages and disadvantages of various cytologic preparatory techniques

Method	Advantages	Disadvantages
Cytocentrifugation	Simple, larger-sized clusters, better preserved architecture	Air-drying artifact; multiple slides need to be prepared due to cell loss; more unsatisfactory or less than optimal specimens
Membrane filter	Good morphology	Difficult to prepare; rapid drying makes storage difficult; cells are distorted by pores; cells are placed in various planes of focus and make screening tedious; background may not be clean; requires fresh specimens as prefixation coagulates proteins which clog the filter; longer screening time
ThinPrep (TP)	Standardized and easy preparation, monolayer, increased cellularity, better preservation, decrease in less than optimal specimens, uniform cell distribution, clean background, shorter and easy screening; multiple slides can be prepared; additional cost is offset by improved specimen quality	Some alteration of key nuclear morphologic features and background elements, fragmentation of cell clusters, cell shrinkage, more expensive than conventional preparations
SurePath (SP)	Standardized preparation, excellent cell yield, preservation of morphology; multiple slides can be prepared; slides are stained on the processor	Cells are in various planes of focus and make screening and focusing at high magnification tedious

robotic arm then sequentially delivers stains to each settling chamber to stain the specimen. The result is a 13-mm circular smear. The PrepStain System processes and stains up to 48 specimens per run.

9. Rinse each slide with absolute alcohol and place in a xylene dish for at least 5 min before coverslipping.

ThinPrep (TP)

Principle

ThinPrep preparation is a filter-based cell concentration technique. Two types are available, TP 2000 and a newer version TP 5000 for use only with TP. Both are FDA-approved for processing Pap test and non-gyn specimens. The TP 2000 processor is a semiautomated device which processes one specimen at a time. TP 5000 is a fully automated benchtop instrument that processes 25 specimens at a time. The latter device requires less hands-on technician time in preparation. Multiple preparations can be made from a single specimen vial. It also has integrated feature of slide etching which reduces time spent in manual application of labels and also reduces potential labeling errors.

Table 1.2 Technical differences between LBP

Features	ThinPrep	SurePath
Cost	Expensive	Less expensive
Sample collection	Uniform	Uniform
Sample transfer	Entire	Entire
Fixation	Immediate	Immediate
Transport	Easy	Easy
Slide preparation	Fully automated	Partial automation
Slide evaluation	Easier	Easy
Cell deposition	Well-defined 20 mm diameter area	Well-defined 13 mm diameter area
Cell preservation	Good	Good
Obscuring factors	None	None
Air-drying	None	None
Screening time	Reduced	Reduced
Reproducibility	Yes	Yes
Ancillary studies	Possible	Possible

Adapted from: Koss LG, Hoda RS. Indication, collection, and laboratory processing of cytologic samples, Chap. 2. In Koss's cytology of the urinary tract with histopathologic correlations. 2012. p. 7–16. With permission of Springer

Table 1.3 General cytologic features on LBP

Features	ThinPrep	SurePath
Quality	Enhanced	Enhanced
Background		
Clean	Yes	Yes
RBCs	Reduced	More reduced
Neutrophils	Reduced	Reduced
Necrosis	Clumped	Clumped, or diffuse
Cellularity	Lower	Higher
Cell distribution	Uniform, one plane of focus	Uniform, thick, different planes of focus
Cell size	Smaller	Small
Architecture	Less well preserved	Better preserved
Cytomorphology	Preserved	Preserved
Extracellular material[a]		
Quantity	Reduced	Less reduced
Appearance	Altered	Altered

Adapted from: Koss LG, Hoda RS. Indication, collection, and laboratory processing of cytologic samples, Chap. 2. In Koss's cytology of the urinary tract with histopathologic correlations. 2012. p. 7–16. With permission of Springer

[a]Extracellular material including necrosis, mucin, and lubricant are altered in quality

Specimen Preservation

The sample is preserved in CytoLyt® solution (methanol-based fixative which is both hemolytic and mucolytic).

Table 1.4 Specific cellular features on LBP

Features	ThinPrep	SurePath
Architecture		
Fragmentation	Present ++	Present +
Monolayer cells	+	–
Cell clusters	Present, 3D, flat, smaller, cohesive, minimal overlap	Present, thick, 3D, > depth of focus, cohesive, more, overlap
Flattening	More	Less
Cellular morphology		
Shape	More rounded	Rounded/elongated
Nucleus		
Detail	Enhanced	Enhanced
Nucleoli	More prominent	Preserved
Inclusions	Preserved	Preserved
Cytoplasm		
Detail	May be denser	May be denser
Shape	Retained	Retained
Elements[a]	Preserved	Preserved

Adapted from: Koss LG, Hoda RS. Indication, collection, and laboratory processing of cytologic samples, Chap. 2. In Koss's cytology of the urinary tract with histopathologic correlations. 2012. p. 7–16. With permission of Springer

+ present, – not present

[a]Cytoplasmic elements include vacuolations, pigment, and PMNs

Slides

The microscopic slides are provided by the company and are marked with a 20-mm diameter circle.

Steps in Preparation

Specimen is first centrifuged at 1500 rpm, and the cell pellet is then resuspended in 30 mL of CytoLyt and again centrifuged. Two to three drops of the cell pellet are transferred to PreservCyt (a methanol-based preservative solution). The vial and a labeled slide are placed into the TP processor. Preparatory steps are similar for both TP 2000 and TP 5000 and include specimen dispersion, collection, transfer, and staining.

1. *Dispersion*: A disposable cylinder with a polycarbonate filter attached to one end is introduced into the vial. The pore size of the filter is 5.5 μm for non-gyn specimens and 8 μm for Pap test specimens. The instrument is rotated creating a current that disaggregates blood, mucus, and other debris and breaks up large cell clusters, mixing and homogenizing the cell suspension.
2. *Collection*: A gentle vacuum is applied to the cylinder which aspirates the cell suspension through the filter. Most of the broken red blood cells (RBCs) and debris are allowed to pass through while the diagnostic cells attach to the external surface of the filter. The instrument monitors cell density across the filter, and

the flow rate decreases when cells are evenly distributed on the filter with minimal overlap.
3. *Transfer*: The cylinder moves out of the specimen, is inverted 180°, and then gently pressed against a positively charged slide and with slight positive pressure, transferring the cells (~70,000) to the glass slide. The result is a 20-mm circular smear with even distribution of cells and minimal overlap. The slide is immediately dropped into 95 % ethanol fixative. Preparation time ranges between 30 and 90 s depending on cell concentration.
4. *Staining*: Papanicolaou (Pap) staining is either performed manually or on an automatic stainer. The staining process takes 30 min. A Papanicolaou stain of fixed samples offers the best option for judging the fine details of cell structure. All alcohol-fixed illustrations in this book are stained using this method.

Residual LBP Specimen

The shelf life of the residual specimen for SP and TP is 3 weeks and 3 months, respectively, at room temperature. The residual specimen can be used for immunochemistry, molecular tests such as UroVysion, or cell block creation. Cellient (Hologic, Bedford, MA) is a new automated cell block machine utilizing centrifugation and filtration, resulting in the capture of cells from a low cellularity specimen such as urinary specimens.

Alterations in General Features in LBP

Main alterations in LBP occur in cellularity, background, architecture, and cellular morphology, induced by collection, fixation, and processing of TP and SP.

Cellularity and Adequacy: For Pap test cytology, an adequate LBP preparation should have an estimated minimum of at least 5000 well-visualized/well-preserved squamous cells. The threshold for minimum cellularity for LBP is lower than the 8,000–12,000 cells for CPS. Guidelines for estimating cellularity of LBP for Pap test cytology are available in The Bethesda System for Reporting Gynecologic Cytology, 2014. Pap test slides with fewer than 5000 cells on LBP should be examined to determine if the reason for the scant cellularity is related to factors such as excessive blood, excessive inflammation, or mucin or whether the problem was indeed low squamous cellularity. Specimens with more than 75 % of squamous cells obscured should be termed unsatisfactory, assuming that no abnormal cells are identified. The median rate for unsatisfactory specimens, from the 2006 practices of participants in the College of American Pathologist (CAP) interlaboratory comparison program for gynecologic cytology, was 1.1 % or less for all preparations. SurePath preparations were associated with the lowest unsatisfactory rate. CPS had the highest 95th percentile rates and TP had the highest median percentile. The most common reason for all unsatisfactory Pap tests was too few squamous cells.

The unsatisfactory rate with TP ranges from 0.3 to 8.3 % and that for SP is 0.23 %. The most frequent causes of unsatisfactory LBP are too few squamous cells, followed by obscuring red blood cells, acute inflammatory cells, and mucin for TP only. The cell enrichment process of SP is capable of handling significantly greater amounts of potentially obscuring blood than membrane filtration method of TP. Pretreatment of TP specimens containing grossly visible blood with glacial acetic acid (GAA) is effective in reducing the unsatisfactory rate, is a cost-effective measure, and may yield additional information. Pang et al. reported that after reprocessing, 68.5 % of the unsatisfactory specimens yielded a satisfactory interpretation which accounted for a reduction of unsatisfactory rate by 18.25 %. In addition significant abnormalities may be detected in the reprocessed slides. Pretreatment of bloody specimens with GAA does not appear to significantly affect high-risk HPV DNA testing in specimens with ASC-US interpretation. Lubricant contamination of Pap test samples may also result in reduced cellularity.

For non-gynecological and fine needle aspiration (FNA), well-defined adequacy criteria have been recommended by "The Paris System for Reporting Urinary Cytology" for urinary samples and by "The Bethesda System for Reporting Thyroid Cytopathology only. Please see respective chapters.

Background: Background is clean and has less or no obscuring elements in LBP as demonstrated in all referenced studies. The TP and SP collection media, PreservCyt, and CytoRich contain mucolytic and hemolytic agents that reduce blood, mucus, polymorphonuclear leucocytes (PMNs), proteinaceous, and necrotic debris. The background changes are more evident in TP. The background elements, particularly, necrosis, may form granular clumps and aggregates and cling to tumor cells (*clinging* diathesis). This altered background may lead to diagnostic problems in LBP.

Cell Distribution: Cells are generally distributed in thin layers with less overlap in LBP, particularly in TP, which consistently produces thinner monolayered preparations. In SP cell clusters appear more three-dimensional. Moreover, in SP, both single cells and clusters are present in different planes of focus which requires constant focus adjustment during slide review. These clumped cell clusters are also more difficult to examine at higher magnification in SP. Cellular alterations in SP are seen because the cells are allowed to settle under the influence of gravity during processing.

Architectural Features: Honeycomb sheets, syncytial fragments, clusters, branching sheets, and papillary formations are all retained in LBP. Fragmentation of large branching sheets and papillary groups is usually seen in both SP and TP. Three-dimensional cellular fragments are flatter in TP and allow for better microscopic assessment. This may be the result of the cell dispersion and cell transfer steps in TP, where the cylinder rotates rapidly in the specimen vial and the positive air pressure applied to the cylinder during transfer of cells to the glass slide, respectively. Despite the flattening effect, three-dimensional appearance of clusters is retained. For SP, balling up of cells with greater depth of focus, smaller strips of cells, complex three-dimensional fragments, more single cells resulting in pronounced dyscohesion, and apparent cellular elongation are also seen. These features probably result from multiple centrifugation steps and because cells are allowed to settle on the slides under the influence of gravity.

Cytology: *In LBP*, cellular and nuclear details are better preserved and enhanced. Although cell shapes are usually preserved, there may be slight cellular elongation in SP. Nucleoli appear prominent and cherry-red in neoplastic lesions and may even be prominent in benign cells. To avoid misinterpretation, it is best to evaluate other nuclear features of malignancy. Intracytoplasmic structures and material are retained including vacuoles, neutrophils, mucus, melanin, glycogen, and other elements.

Advantages of Liquid-Based Preparations

- Uniform collection technique with theoretical collection of 100 % of the sample.
- Standardized processing techniques produce a homogenized sample with uniform cell distribution with minimal thick areas and cellular overlap.
- TP preparation technique is simple and less labor intensive.
- Reproducibility with less variability in specimen preservation, staining, and quality.
- Less cell loss compared to cytospins.
- Easier and less time-consuming to screen due to smaller and well-defined screening area (TP, 20 mm, and SP, 13 mm).
- Easier and less time-consuming to screen due to cleaner background with minimal or absent obscuring elements.
- SP collection material is also more effective in reducing blood compared with TP.
- Increased cell yield, even with low cellularity samples.
- Better cell preservation due to immediate liquid fixation with no air-drying artifact.
- Other artifacts of smearing such as crush artifact, cell overlap, and thick areas are usually lacking.
- Easier to interpret due to less or no obscuring elements and less cellular overlap.
- Enhanced cellular and nuclear morphology resulting in high diagnostic accuracy.
- Cytologic features of malignancy are retained.
- Decreased rate of inadequate or unsatisfactory specimens makes the LBP more cost-effective.
- Preferred preparatory method at FNA procedures when a cytologist is not present.
- Residual specimen vial can be used to prepare additional slides or utilized for ancillary tests.
- In Pap test cytology, processed as TP or SP, automated computerized screening devices with programmed algorithms locate and mark worrisome cells making screening more accurate and increasing productivity.

Disadvantages of LBP

These are predominantly due to preparation techniques.

- Fragmentation of papillae and cell groups and slightly more dyscohesion of cells may be more pronounced in TP.
- Background material may be lost, reduced, or altered and may be more pronounced in TP.
- In TP, excess blood may remain.
- In SP, multiple processing steps and settling of cells under the influence of gravity result in cellular elongation, cells in different planes of focus, and increased three-dimensional appearance.
- Cell and nuclear size may become smaller in both TP and SP.
- Immediate adequacy assessment cannot be performed at FNA procedures.
- The need for ancillary studies such as flow cytometry and microbiology culture cannot be assessed at FNA procedures.

In summary, liquid-based preparations (LBP) have largely replaced conventional Pap smears for cervical samples in the USA and in many other industrialized countries. Split sample and direct-to-vial studies have shown that LBP show an overall improvement in collection and processing of samples, reduce artifacts, and can be utilized for HPV test, and diagnostic accuracy of LBP is comparable to conventional Pap smears. Additional advantages of LBP for Pap test in comparison to conventional Pap smears include standardized automated preparations, automated screening, improved specimen adequacy, reduced unsatisfactory rate, and ability to perform HPV test. Advantages of LBP make them cost-effective.

Liquid-based preparations are increasingly being used for non-gynecologic cytology including FNA, for which conventional smears were standard diagnostic preparatory method. The advantages listed above for Pap test cytology also apply to non-gynecologic specimens. In FNA the LBP are mostly being used in conjunction with direct smears. However, there is a growing trend to process the entire FNA sample as one LBP with/without a cell block.

Cytologists need to be aware of cytological features on LBP as outlined in this book.

Suggested Reading

1. Faquin WC. Use of conventional smears versus liquid-based preparations for fine-needle aspirate specimens: the time has not come to abandon your conventional smears! Cancer Cytopathol. 2014;122:340–2.
2. Fontaine D, Narine N, Naugler C. Unsatisfactory rates vary between cervical cytology samples prepared using ThinPrep and SurePath platforms: a review and meta-analysis. BMJ Open. 2012;13(2):e000847.
3. Lin SN, Taylor J, Alperstein S, Hoda R, Holcomb K. Does speculum lubricant affect liquid-based Papanicolaou test adequacy? Cancer Cytopathol. 2014;122(3):221–6.

4. Hoda RS, Loukeris K, Abdul-Karim FW. Gynecologic cytology on conventional and liquid-based preparations: A comprehensive review of similarities and differences. Diagn Cytopathol. 2013;41:257–78.
5. Hoda RS. Non-gynecologic cytology on liquid-based preparations: a morphologic review of facts and artifacts. Diagn Cytopathol. 2007;35:621–34. Review
6. Ha SY, Lee YK, Oh YL. Effectiveness of the ThinPrep imaging system in the detection of abnormal cervicovaginal cytology: a practical experience in Korea. Acta Cytol. 2013;57:159–63.
7. Thrall MJ, Russell DK, Bonfiglio TA, Hoda RS. Use of the ThinPrep imaging system does not alter the frequency of interpreting Papanicolaou tests as atypical squamous cells of undetermined significance. CytoJournal. 2008;5:10.
8. Wilbur DC, Black-Schaffer WS, Luff RD, Abraham KP, Kemper C, Molina JT, Tench WD. The Becton Dickinson FocalPoint GS imaging system: clinical trials demonstrate significantly improved sensitivity for the detection of important cervical lesions. Am J Clin Pathol. 2009;132:767–75.
9. Cibas ES, Ducatman B. Cytology: diagnostic principles and clinical correlates. 4th ed. Philadelphia: Saunders Elsevier; 2014.
10. Elsheikh TM, Kirkpatrick JL, Wu HH. Comparison of ThinPrep and cytospin preparations in the evaluation of exfoliative cytology specimens. Cancer. 2006;108:144–9.
11. Nayar R, Wilbur DC, editors. The Bethesda system for reporting cervical cytology: definitions, criteria, and explanatory notes. 3rd ed. Cham: Springer; 2015.
12. Rosenthal DL, Wojcik EM, Kurtycz D. The Paris system for reporting urinary cytology. New York: Springer; 2016.
13. Renshaw AA, Elsheikh TM. Assessment of manual workload limits in gynecologic cytology: reconciling data from 3 major prospective trials of automated screening devices. Am J Clin Pathol. 2013;139:428–33.
14. Tripathy K, Misra A, Ghosh JK. Efficacy of liquid-based cytology versus conventional smears in FNA samples. J Cytol. 2015;32:17–20.
15. Tabatabai ZL, Auger M, Kurtycz DF, Laucirica R, Souers RJ, Nayar R, Khalbuss WE, Moriarty AT, Fraig M. Do liquid-based preparations of pulmonary bronchial brushing specimens perform differently from classically prepared cases for the diagnosis of malignancies? Observations from the College of American Pathologists interlaboratory comparison program in nongynecologic cytology. Arch Pathol Lab Med. 2015;139:178–83.
16. Randolph ML, Wu HH, Crabtree WN. Reprocessing unsatisfactory ThinPrep papanicolaou tests using a modified SurePath preparation technique. Cancer Cytopathol. 2014;122:343–8.
17. Malapelle U, de Rosa N, Rocco D, Bellevicine C, Crispino C, Illiano A, Piantedosi FV, Nappi O, Troncone G. EGFR and KRAS mutations detection on lung cancer liquid-based cytology: a pilot study. J Clin Pathol. 2012;65:87–91.
18. Chang H, Lee H, Yoon SO, Kim H, Kim A, Kim BH. BRAF(V600E) mutation analysis of liquid-based preparation-processed fine needle aspiration sample improves the diagnostic rate of papillary thyroid carcinoma. Hum Pathol. 2012;43:89–95.
19. Ren S, Solomides C, Draganova-Tacheva R, Bibbo M. Overview of nongynecological samples prepared with liquid-based cytology medium. Acta Cytol. 2014;58:522–32.

Gynecologic Cytology

2

Introduction

Cytological screening for cervical carcinoma and its precursor lesions using the Papanicolaou smear test (Pap test) has been efficient in reducing its morbidity and mortality. Currently, cervical cancer is the seventh leading cause of cancer deaths in the United States. In 2016, about 12,340 new cases of invasive cervical cancer will be diagnosed, and about 4030 women will die from the disease. While the conventional Pap smear (CPS) has been successful in reducing the incidence and mortality rate from cervical cancer, it has limitations, particularly with respect to false-negative screening results. Several properties of CPS lead to high false-negative results such as: In CPS, the collected cell sample is smeared on the slides which often results in thick cellular areas making the finding of an abnormality difficult. Other background elements such as blood, inflammatory cells, and necrosis may also obscure diagnostic cells; and artifacts such as air-drying and crush effect may further compromise the specimen. The liquid-based preparations (LBPs) have revolutionized the Pap test by allowing standardization of cervical specimen collection, processing, and screening. LBP has shown a significant improvement in the detection of precursor lesions of cervical neoplasia. In LBP, the cells are rinsed into a liquid preservative collection medium [CytoLyt for ThinPrep (TP) and BD® SurePath (SP) preservative for SP specimens]. This immediate wet fixation of specimen ensures good preservation and reduces the time for air-drying artifact to be introduced. Specimen processing is automated and, thus, standardized and uniform. TP specimens are processed using TP2000 or TP5000 processors, and SP specimens are processed using PrepMate or Totalys System. Residual specimens left in collection media can be used for ancillary testing of human papillomavirus (HPV), chlamydia, and gonorrhea. Despite the differences in preparatory techniques, the two cytological preparations using LBP (TP and SP) are largely similar; however, subtle differences exist. The Bethesda System (TBS) for Reporting Cervical Cytology, updated in 2014, provides criteria for Pap test diagnosis.

R.S. Hoda et al., *Diagnostic Liquid-Based Cytology*,
DOI 10.1007/978-3-662-53905-7_2, © Springer-Verlag GmbH Germany 2017

Specimen Procurement and Fixation

Please see Chap. 1.

Alterations in General Features in LBP

The main alterations occur in background, architecture, and cellular morphology. These alterations are probably due to technical reasons (Please see Chap. 1).

The 2014 Bethesda System for Reporting Cervical Cytology

An update to TBS was released in 2014. This update allowed improved clarifications to the guidelines, as well as the addition of images [1]. In contrast to the previous guidelines, the 2014 guidelines recommend that the "other" category of benign-appearing endometrial cells be reported in women age of 45 or greater (previously, the age cutoff was more than or equal to 40). The guidelines also address the use of a "low-grade squamous intraepithelial lesion/cannot exclude high-grade squamous intraepithelial lesion" (LSIL-H) diagnostic category; this category has not been accepted by TBS, and the 2014 guidelines recommend against its use.

American Society for Colposcopy and Cervical Pathology (ASCCP) Interim Guidelines for Primary High-Risk HPV (hrHPV) Testing

In 2011, the American Society for Colposcopy and Cervical Pathology (ASCCP) released updated consensus guidelines for the management of abnormal Pap tests [3]. Following approval of HPV testing as a primary screening test for cervical dysplasia by the Food and Drug Administration (FDA) in 2014, the ASCCP released interim guidelines for primary HPV testing. While cytology alone and cotesting are the only screening options specifically recommended in major guidelines, primary high-risk HPV (hrHPV) screening can be considered as an alternative. While data is still limited, the interim guidelines suggest that hrHPV-positive women should have HPV genotyping performed, with HPV 16/18-positive women being triaged to colposcopy and women positive for the other 12 hrHPV genotypes triaged to cytology. With the latter triage, a diagnosis of ASC-US or higher on the Pap test results should be sent to colposcopy. Primary hrHPV screening should not be performed in women below the age of 25.

HPV Vaccination

Three HPV vaccines have been approved by the FDA: Gardasil, Gardasil 9, and Cervarix. All three vaccines protect against HPV types 16 and 18, which cause approximately 2/3 of all cervical cancers. Gardasil also prevents HPV types 6 and 11, which cause the majority of genital condylomas, and Gardasil 9 protects against five additional high-risk HPV types (31, 33, 45, 52, and 58). Females and males between the ages 9 and 26 are eligible for Gardasil, except Gardasil 9 is only approved for males between the ages 9 and 15. Cervarix is approved for use in females between 9 and 25 years old. Studies have shown protection to last for at least 8–9 years. However, given the relatively recent use of these vaccines, their long-term effects remain unknown; therefore, it is difficult to predict how widespread HPV vaccination will impact HPV prevalence and future screening guidelines for cervical cancer and dysplasia.

Immunocytochemistry (ICC) on LBP

Several studies have reported on the utility of ICC in LBP. Immunostaining on LBP shows equal or greater intensity and proper distribution of staining compared to CPS. Good results have been reported for epithelial, lymphoid, neuroendocrine, sarcoma, melanoma, molecular, and prognostic markers. However, ICC on cell block (CB) gives superior results compared to LBP.

Assimilation of HPV oncogenes E6 and E7 into the host DNA promotes upregulation of cyclin-dependent kinase inhibitor (CDKI) *p16* (INK4A). The latter is detectable by monoclonal antibody in the developing cervical cancer cells. p16 immunostaining has been successfully applied to LBP for both squamous and glandular lesions. An association has been shown between strong p16INK4A immunostaining of atypical squamous/glandular cells in smears and the presence of a significant lesion in the cervix. This antibody is not expressed in normal glandular cells. p63 stains basal cells and may be a diagnostic pitfall in atrophic Pap tests.

Automation

One benefit of LBP is the automation used for both specimen preparation as well as digital screening of specimens. This is important since Pap tests remain the largest specimen type in most cytopathology laboratories, and automation decreases laboratory technician, cytotechnologist, and cytopathologist workload, which in turn allows for increased productivity. For instance, the ThinPrep TP5000 processor allows continuous, hands-free processing of 20 Pap and non-Gyn specimens. Coupled with an autoloader, the system can allow for up to 8 h of hands-free technician time. Multiple automated systems for the digital screening of Pap tests exist, such as the ThinPrep Imaging System (TIS, Hologic Corp., Marlborough, MA) for use on TP and BD FocalPoint Slide Profiler and BD FocalPoint GS™ Imaging System (BD Diagnostic, Burlington, NC) for use in SP and CPS. These systems allow for Pap tests to be digitally screened.

Conclusion

In conclusion, LBP are of great value for the processing of gynecologic specimens (Figs. 2.1–2.28). Most diagnostic criteria utilized on CPS are also applicable to LBP. The seitivity and specificity of LBP are either comparable or superior to CPS in detecting abnormalities. The unsatisfactory rate is lower in comparison o CPS. LBP are less time-consuming to screen and easier to interpret, as the cells are limited to a smaller area on a cleaner background with excellent cellular preservation. The use of LBP has the potential to decrease turnaround time and increase the number of specimens being tested in a laboratory without corresponding increase in technical staff, a quality which also makes LBP more cost-effective. LBP requires familiarity with alterations in background, extracellular elements, architecture, and cell morphology. Experience (and in some situations, modification to diagnostic criteria previously utilized for CPS) may be required.

The 2014 Bethesda System for Reporting Cervical Cytology

- Negative for intraepithelial lesion or malignancy
  - Nonneoplastic findings, organisms
- Other
- Epithelial cell abnormalities
  - Squamous cell
    - Atypical squamous cells
      Of undetermined significance (ASC-US)
      Cannot exclude HSIL (ASC-H)
    - Low-grade squamous intraepithelial lesion (LSIL)
    - High-grade squamous intraepithelial lesion (HSIL)
    - Squamous cell carcinoma (SqCC)
  - Glandular cell
    - Atypical, not otherwise specified (NOS)
      - Endocervical cells, endometrial cells, glandular cells
    - Atypical
      - Endocervical cells, favoring neoplasia (AEC-FN)
      - Glandular cells, favoring neoplasia (AGC-FN)
    - Endocervical adenocarcinoma in situ (AIS)
    - Adenocarcinoma
      - Endocervical, endometrial, extrauterine, and NOS
    - Other malignant neoplasms

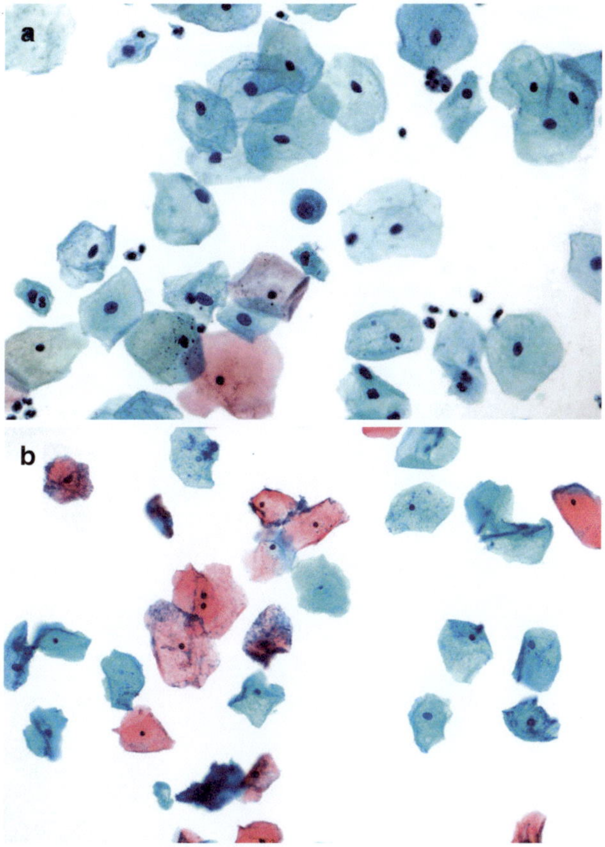

Fig. 2.1 Benign squamous cells. (**a, b**) Superficial cells are flat with abundant, typically eosinophilic cytoplasm and pyknotic nuclei. Intermediate cells are also flattened with abundant basophilic or eosinophilic cytoplasm and vesicular nuclei. Parabasal cells are round and have basophilic, dense cytoplasm; the nucleus is larger and typically round or oval in shape; and small nucleoli are often present (**a**, TP; **b**, SP)

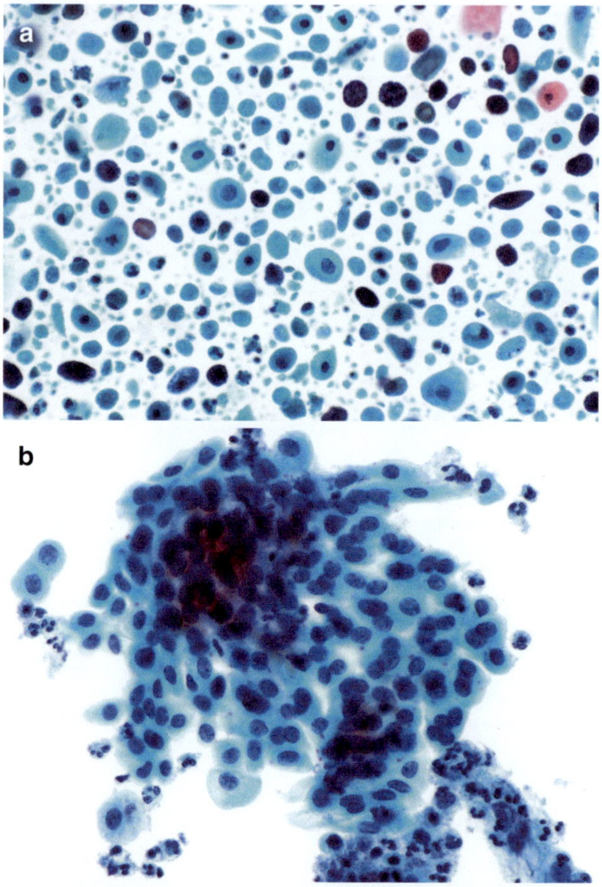

Fig. 2.2 Atrophy. (**a**) Parabasal cells with mostly bland nuclei and some with eosinophilic cytoplasm are seen. Interspersed degenerated cells with pyknotic nuclei are also present. Background shows clumped basophilic granular debris and inflammation (SP). (**b**) Atrophy shows a cluster of parabasal and basal cells. Parabasal cells, seen toward the periphery, are rounder with dense well-defined cytoplasm, round slightly dark, uniform nuclei and small nucleoli. Nuclear to cytoplasmic ratio is low. Basal cells have oval nuclei and less cytoplasm with a relatively high nuclear to cytoplasmic ratio. Cells with eosinophilic cytoplasm are lacking (TP). An atrophic picture can result from many different clinical etiologies, most of which are associated with decreased estrogen (e.g., postmenopause or postpartum). Cells with high nuclear to cytoplasmic ratios may cause suspicion for high-grade squamous intraepithelial lesion (HSIL). In atrophic vaginitis, the background shows clumps of granular inflammatory debris, similar to the diathesis seen in squamous cell carcinoma and distinguished from it only by the absence of malignant cells in atrophy. Inflammation does not obscure cell detail in LBP

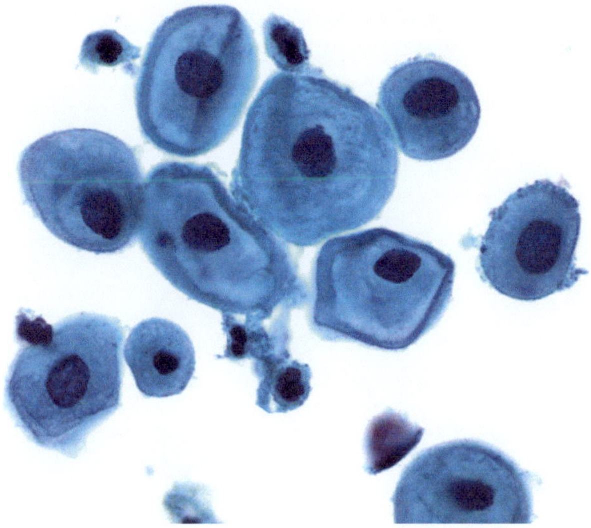

Fig. 2.3 Navicular cells. Navicular cells are intermediate squamous cells containing glycogen. In LBP, glycogen may be lost during processing and appear as an equivocal cytoplasmic halo with a thick rim, which if visualized carefully appears to be a cytoplasmic fold. The nuclei are uniform, small, and vesicular. Low-grade squamous intraepithelial lesion (LSIL) may be mistaken for navicular cells and vice versa. However, normal intermediate cell nuclei are helpful in avoiding overinterpretation (TP)

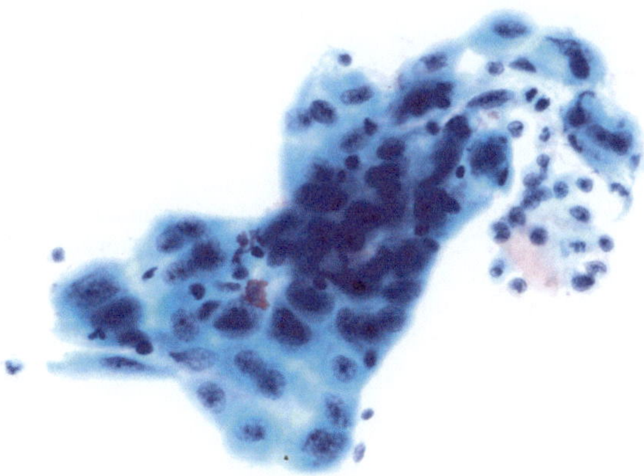

Fig. 2.4 Repair. In contrast to conventional smears, cells with repair changes appear as cohesive monolayer sheets of cells with less "streaming" effect and rounded, instead of frayed, cell borders. The staining of cells may also be more uniform with less polychromasia. Cells may be bi- or multinucleated with enlarged, round to oval, uniform, and regular nuclei and evenly dispersed pale chromatin. Macronucleoli are the most significant feature; cytoplasm may be vacuolated with intracytoplasmic neutrophils. Nuclear to cytoplasmic ratio is low and background may show inflammation (TP). The differential diagnosis is similar to conventional smear and includes invasive squamous cell carcinoma. In the College of American Pathologists (CAP) Interlaboratory Comparison Program in Gynecologic Cytology, TP slides with a reference diagnosis of reparative change had a lower false-positive and discordance rate and a higher exact match error rate than CPS

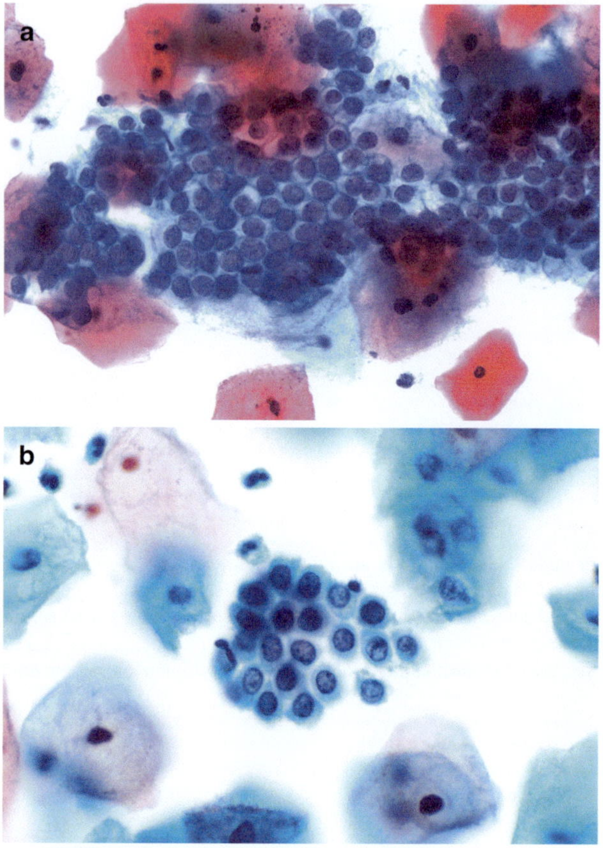

Fig. 2.5 Benign glandular endocervical cells. (**a, b**) Endocervical cells are present and evenly spaced in a honeycomb sheet with retained polarity and no cellular overlap. The cytoplasm is mucinous, and nuclei are round and uniform with pale chromatin, and chromocenters with occasional protrusion ("nippling"). Benign endocervical cells appear similar in conventional smears and LBP. However, individual cells may be tall and columnar, a feature that is more pronounced in SP. In SP, nuclear contour is regular. The cytoplasm is often dense with small vacuoles and distinct cell borders, and the nuclear to cytoplasmic ratio may be slightly increased (**a**, TP; **b**, SP)

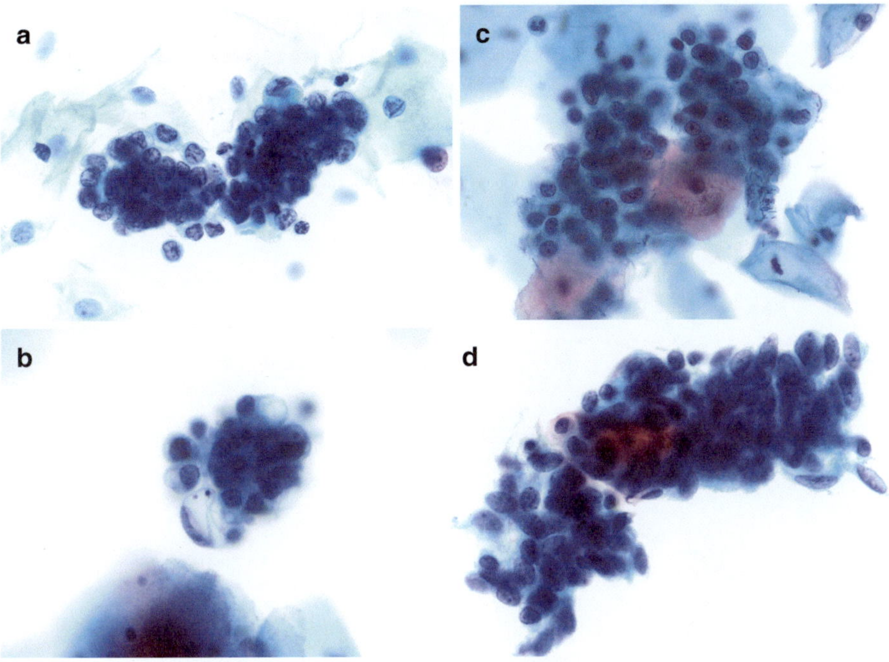

Fig. 2.6 Benign endometrial cells. (**a, b**) Endometrial cells are arranged as a three-dimensional group with scalloped borders. The nuclei are round and small (equal to intermediate cell nuclei) and possess small nucleoli or chromocenters. The cytoplasm is vacuolated (**a**, TP; **b**, SP). Exodus as seen in (**a**) appears as a double-contoured round to oval cluster of epithelial cells surrounding a dense core of stromal cells. Apoptosis may be present. (**c**) Superficial stromal cells are round cells with round pale nuclei and foamy cytoplasm. These cells tend to stay together as a loosely cohesive group. (**d**) Deep stromal cells are elongated with elongated, spindled, and dark nuclei and scant bipolar cytoplasm. In CPS, the cytoplasm is not usually evident (**c**, **d**, TP). The differential diagnosis of deep stromal cell includes HSIL. Immunostain for p16 may be helpful as it would be positive in HSIL, while CD10 would be positive in deep stromal cells

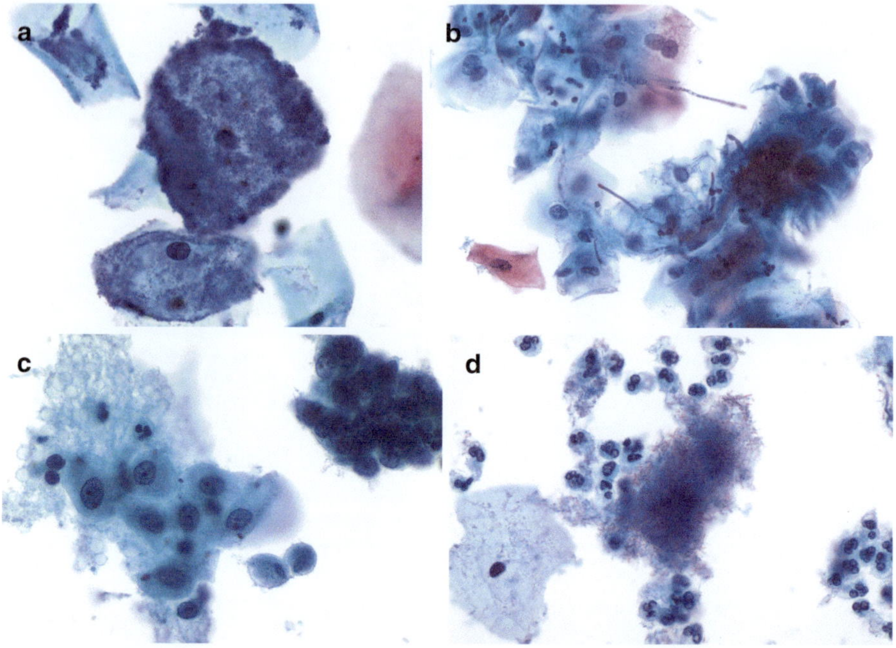

Fig. 2.7 Bacterial microorganisms. (**a**) "Shift in vaginal flora" or bacterial vaginosis. The pathognomonic "clue cells" are intermediate squamous cells heavily covered with bacteria, giving a "shag carpet" appearance. The background is usually clean with clumps or loose clusters of organisms and is devoid of inflammatory cells (TP). (**b**) *Candida* sp. pseudohyphae appear as long basophilic to pinkish structures skewering the superficial squamous cells in a "shish kebab" appearance; yeast forms may also be present (TP). (**c**) *Trichomonas vaginalis*. The organisms appear small and pear shaped (occasionally "kite shaped") with an eccentric single almond-shaped basophilic nucleus, cherry-red or basophilic granules, and one or more flagella. The organisms may overlie squamous cells, particularly in SP. Although the clean background of LBP makes identification of *Trichomonas* less laborious, the smaller-sized trophozoites may be eliminated during processing. Degenerating parabasal and endocervical cell cytoplasmic fragments may mimic *Trichomonas*. Squamous cells show reactive changes (TP). (**d**) *Actinomyces*. The organism appears as gray-blue to black-staining small islands of dense amorphous material. Closer examination reveals tangled clusters and haphazardly arranged filamentous bacterial colonies which branch at acute angles and often contain blunt ends. Several loose organisms are also seen admixed with inflammatory cells (TP)

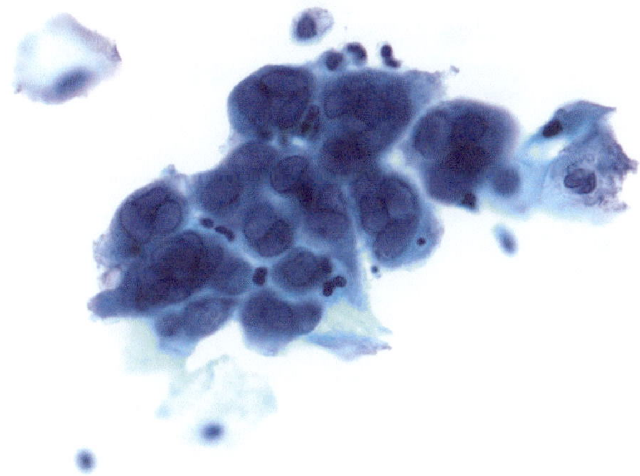

Fig. 2.8 Herpes simplex virus. The morphology is similar to conventional smear except that the virus is easier to detect in LBP due to the clean background. Note the multinucleation and molding of ground-glass nuclei, the latter consisting of the actual viral particles. Cytoplasmic inclusions may also be seen (TP)

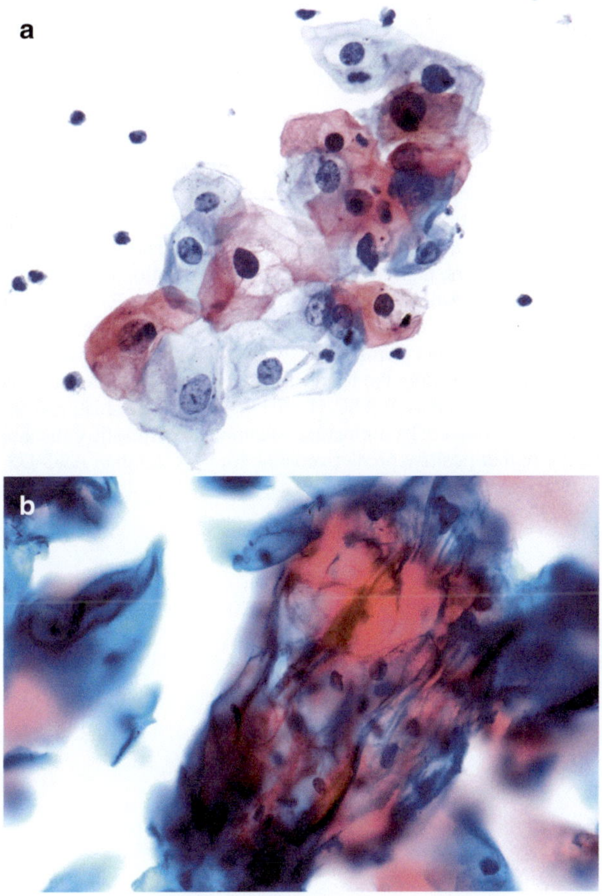

Fig. 2.9 Atypical squamous cells of undetermined significance (ASC-US). (**a**, **b**) Mature squamous cells, in TP, show nuclear features of slight enlargement, subtle membrane irregularities, and chromatin changes. In both TP and SP, cytoplasm shows suggestion of koilocytotic halos (**a**, TP; **b**, SP). The cytological criteria for ASC-US in conventional smears and LBP are similar. As a result of the improved ability to discern benign mimickers of ASC-US, LBP has shown a decrease in the interpretation of ASC-US, and this diagnosis is associated with an increased likelihood of representing a significant lesion. In LBP, ASC-US cells are more likely to be dispersed

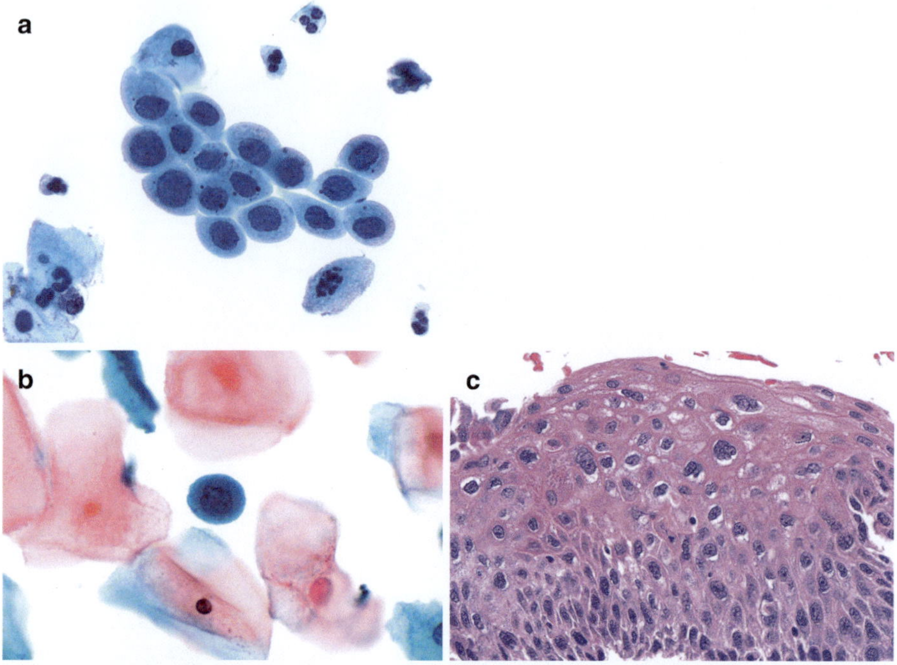

Fig. 2.10 Atypical squamous cells, cannot exclude HSIL (ASC-H). (**a**) In this group of ASC-H cells, the nuclei are 1.5–2.5× that of normal metaplastic cells, with slight increase in nuclear to cytoplasmic ratio. The cells are slightly hyperchromatic with slight nuclear irregularity (TP). (**b**) A single ASC-H cell is in the center of the field and demonstrates a hyperchromatic nucleus with irregular borders, high nuclear to cytoplasmic ratio, and dense, basophilic cytoplasm (SP). (**c**) On follow-up, the patient was found to have HSIL (H&E). ASC-H comprises approximately 5–10 % of all ASC diagnoses. An interpretation of ASC-H is generally reserved for few atypical immature metaplastic cells occurring singly or in small loose groups. The reproducibility of ASC-H diagnosis is poor, suggesting the need for more refined criteria and/or continuing education in LBP. The differential diagnosis of ASC-H cells includes HSIL, histiocytes, exfoliated endometrial epithelial and deep stromal cells, immature squamous metaplastic cells, and ASC-US. ASC-H has a significantly higher positive predictive value for CIN 2/3 than ASC-US (50 vs. 17 %)

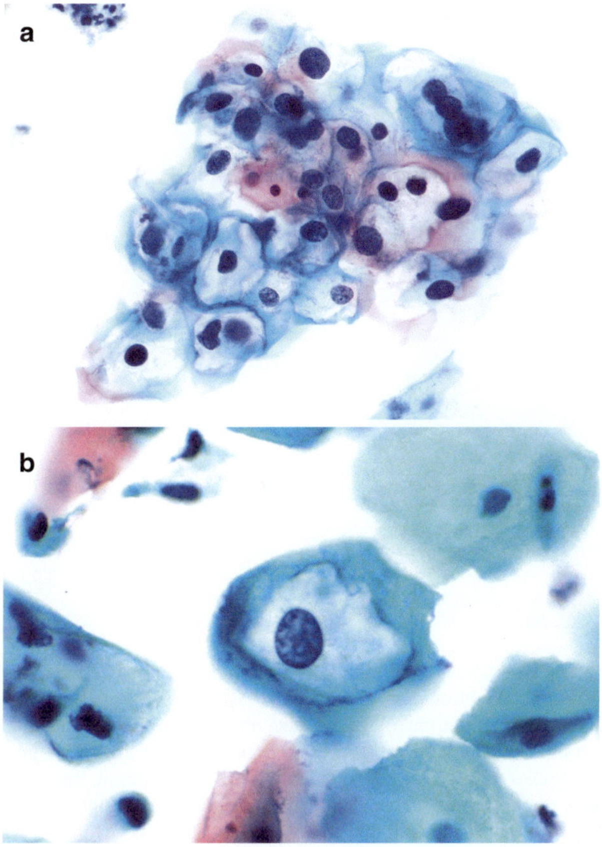

Fig. 2.11 Low-grade squamous intraepithelial lesion (LSIL). (**a**, **b**) In LBP samples, the criteria for LSIL are the same as for conventional smears. LSIL shows increased nuclear detail including chromatin texture, irregularity of membranes, and koilocytotic cavities with well-defined, sharp perinuclear halo and dense peripheral cytoplasmic condensation. Note nuclear hyperchromasia with uniformly distributed chromatin. Contrast this to the findings in navicular cells in Fig. 2.3 in which the cytoplasmic cavity is variable, is not well defined, and may show yellow-staining glycogen and normal nuclei and with ASC-US in Fig. 2.9a, b in which halos around the enlarged nuclei are not well defined and characteristic nuclear features of LSIL are not present (**a**, TP; **b**, SP)

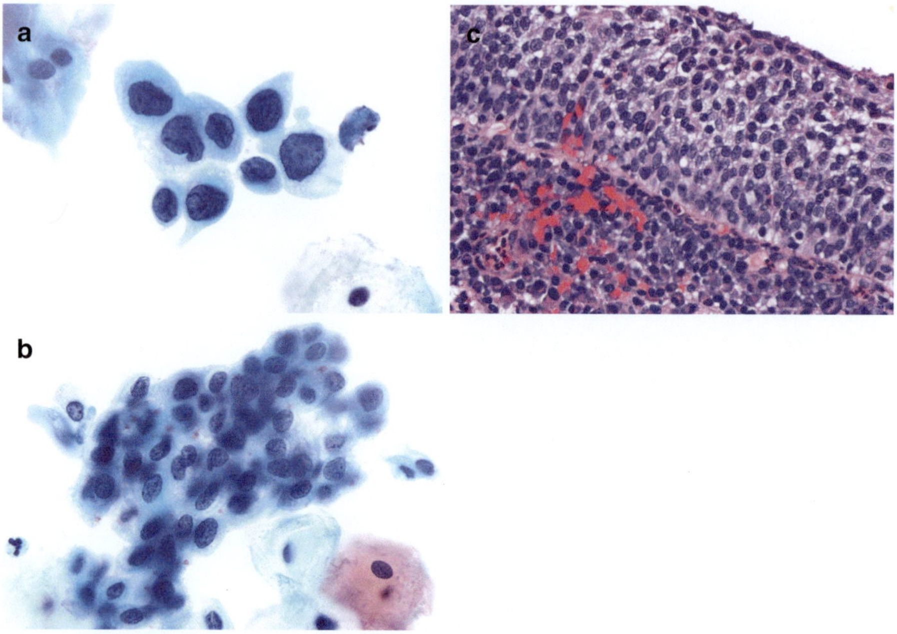

Fig. 2.12 High-grade squamous intraepithelial lesion (HSIL). (**a, b**) While morphologic changes are generally similar in conventional smears and LBP, HSIL cells, in LBP, may appear somewhat smaller than conventional counterpart. While sheets and syncytial groupings are generally maintained, more abnormal cells are isolated or present in small rather than large groups. Note hyperchromatic irregular nuclei with coarse uniformly distributed chromatin, thick nuclear membranes and grooves, small nucleoli, and scant dense metaplastic-type cytoplasm (**a** and **b**, TP). (**c**) Follow-up tissue specimen demonstrating HSIL which also focally involved endocervical glands (H&E)

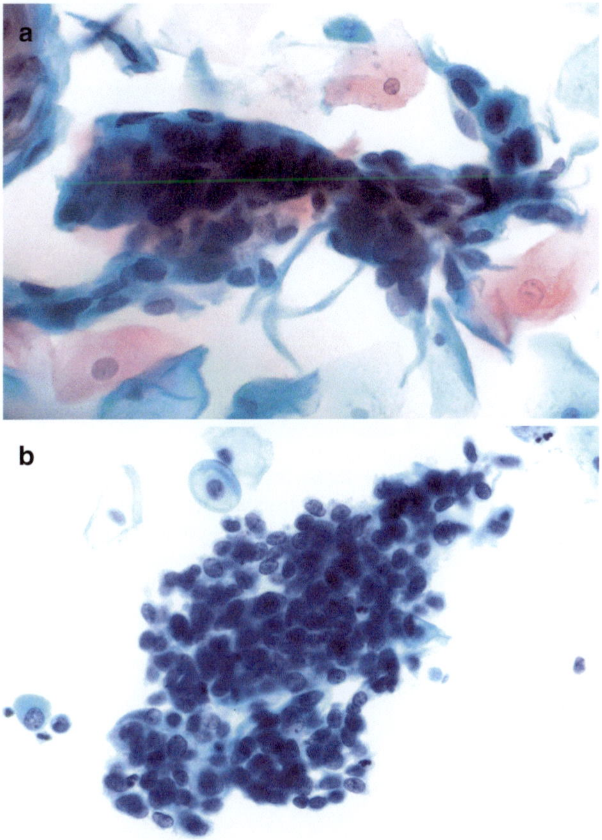

Fig. 2.13 HSIL involving endocervical glands. (**a**, **b**) Such cases demonstrate somewhat densely packed cell groups with loss of central cell polarity and piling within cell groups, a finding not present in endocervical adenocarcinoma in situ (AIS). Central cell polarity is maintained in cellular groupings of AIS. Also, note peripheral flattening of cells, oval or angulated hyperchromatic nuclei, and micronucleoli (**a**, SP; **b**, TP). HSIL involving endocervical glands can mimic endocervical glandular neoplasia. It is important to cytologically distinguish between the two as colposcopic examination may be negative in such lesions and treatment options may vary. Cytologic features of HSIL involving endocervical glands on LBP are similar to those previously reported on conventional smear

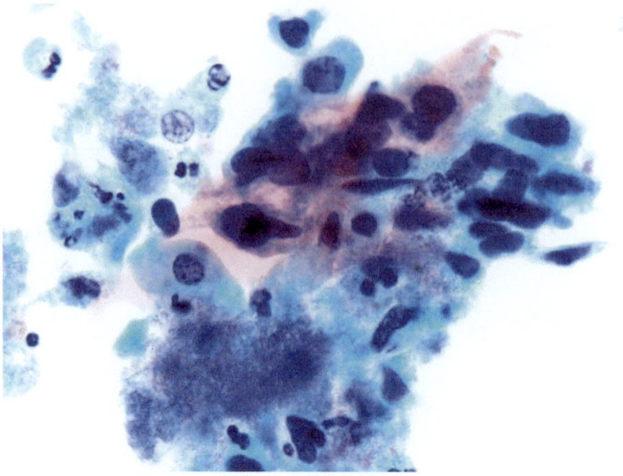

Fig. 2.14 Tumor diathesis. In LBP, tumor diathesis is clumped or clings to tumor cells ("clinging" diathesis). Note in this case the clinging diathesis from a case of keratinizing squamous carcinoma. The background is otherwise clean, in contrast to that in conventional smears. The presence of diathesis-like material should make one suspicious for an invasive carcinoma, though other materials (such as degenerated blood from the endometrium) can mimic tumor diathesis. Metastatic lesions often lack diathesis (TP)

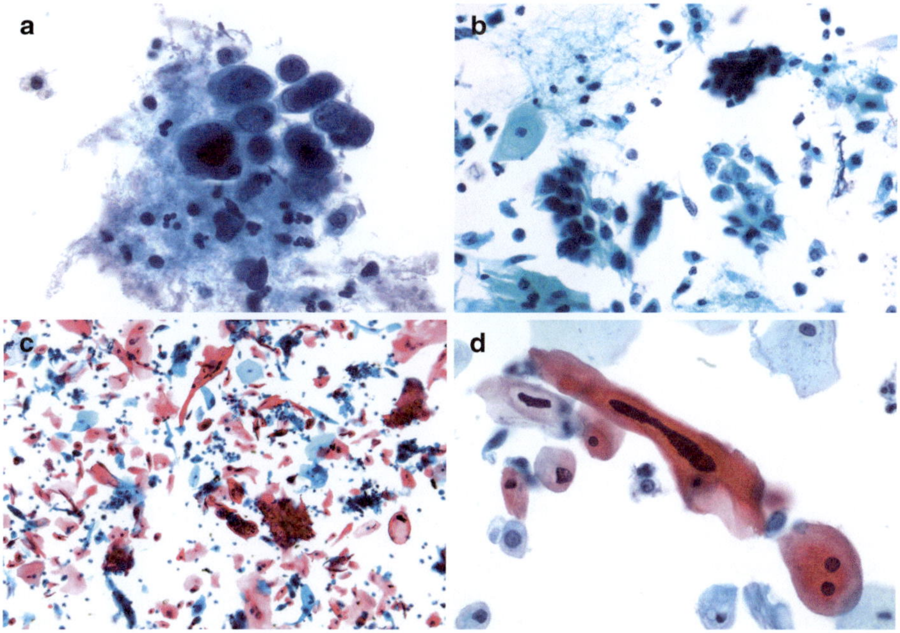

Fig. 2.15 Squamous cell carcinoma. (**a**, **b**) Nonkeratinizing squamous cell carcinoma shows discohesive abnormal cells with some aggregation. There is loss of cellular and nuclear polarity. Nuclei are mostly round and hyperchromatic with a prominent nucleolus and parachromatin clearing. The nuclear to cytoplasmic ratio is high. Note distinctive granular-necrotic diathesis "clinging" to malignant cells. Unlike conventional smears, the "clinging diathesis" does not obscure cell detail (**a**, TP; **b**, SP). In LBP, nonkeratinizing squamous carcinoma cells have a greater depth of focus than conventional smears and are considered in the differential diagnosis of hyperchromatic crowded groups (HCG). Tumor diathesis is prominent in the background and there may also be coexistent HSIL. (**c, d**) Keratinizing squamous cell carcinoma shows many cells with irregularly shaped eosinophilic to orangeophilic dense cytoplasm and dark (almost black) nuclei. There is pleomorphism of both the nucleus and cytoplasm. The latter shows "tadpole and fiber cell" conformations (**c**, SP; **d**, TP)

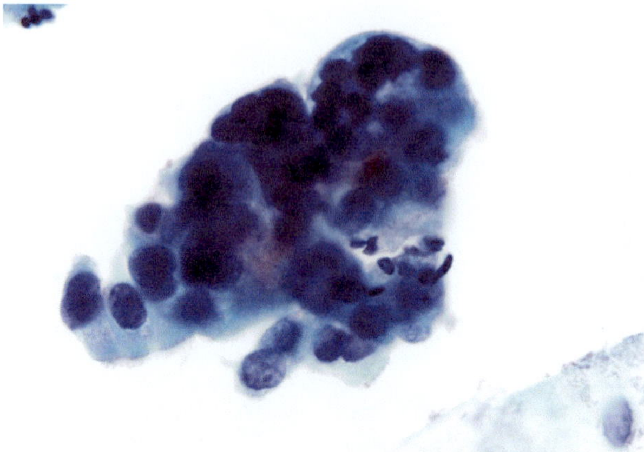

Fig. 2.16 Tubal metaplasia shows ciliated, non-mucin-secreting columnar cells and is considered a normal component of the endocervical epithelium, most common in upper endocervical canal and lower uterine segment (LUS). In LBP, tubal metaplasia occurs as tight, thick, three-dimensional HCGs with a smooth community border and stratified elongated, hyperchromatic nuclei with nucleoli. Cilia can be seen as eosinophilic brush border, particularly in TP; occasionally, cilia are lost during processing, but terminal bars are usually retained. Nuclei of tubal metaplastic cells are centrally placed, slightly enlarged, oval, and sometimes hyperchromatic with clumsy, haphazard arrangement or pseudofeathering; the cytoplasm may be foamy. Multinucleation and nucleoli may be present; thus, tubal metaplasia can mimic AIS (TP)

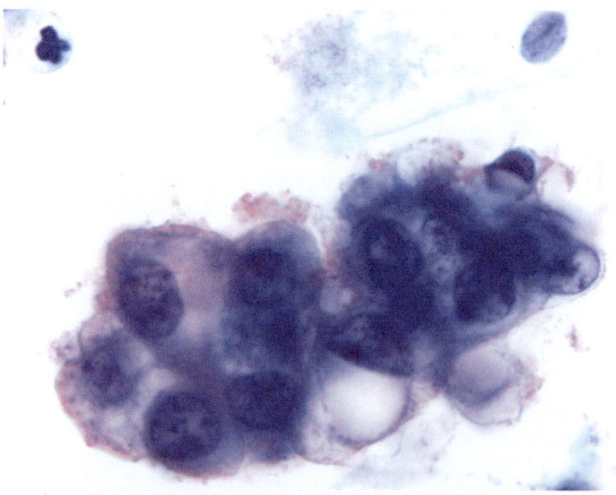

Fig. 2.17 Nonneoplastic glandular changes. In patients with intrauterine contraceptive device (IUD), cellular changes can be seen in both endometrial and endocervical cells. The cells are large and have smooth nuclei with smudged/glassy chromatin and prominent chromocenters. Occasional nuclear clefts are also seen. The cytoplasm shows large degenerative vacuoles (seen here) that push the nucleus toward the periphery. IUD cells are much more easily recognized in LBP due to cleaner background. Clustered IUD cells may mimic well-differentiated endometrial adenocarcinoma, and isolated IUD cells may mimic HSIL. IUD cell cluster is distinguished from the former by age and the presence of an IUD. IUD cells can be distinguished from HSIL as there are rare IUD cells and by clinical history of IUD. Human papillomavirus (HPV) test is usually positive in HSIL (TP)

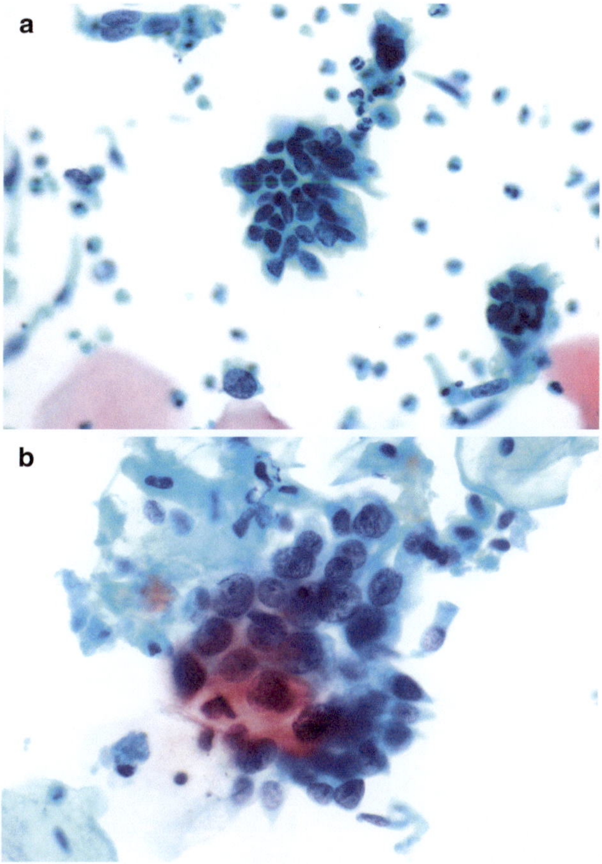

Fig. 2.18 (**a**) Atypical endocervical cells, not otherwise specified (AEC-NOS). Cells are arranged in a two-dimensional sheet. Nuclei are mildly atypical and show focal peripheral palisading or "feathering" with inconspicuous nucleoli (SP). Likewise, the category of atypical endometrial cells, also shows mild cytological atypia. The category of AGC-NOS shows mild cytological atypia, and it may be difficult to determine if these cells are endocervical or endometrial. (**b**) Atypical endocervical cells, favor neoplastic (AEC-FN). The endocervical cell group shows loss of polarity and focal nuclear peripheral palisading or "feathering." The nuclei are slightly irregular with stippled chromatin and contain nucleoli. Cytoplasm is scant (TP). AEC-FN category is reserved for cases in which the cellular changes are more pronounced than those seen in AEC-NOS, but lack the diagnostic features of AIS and endocervical adenocarcinoma. Detection of biopsy-proven glandular neoplasia/AIS is improved with TP compared to conventional smears (14.8 vs. 2.1 %)

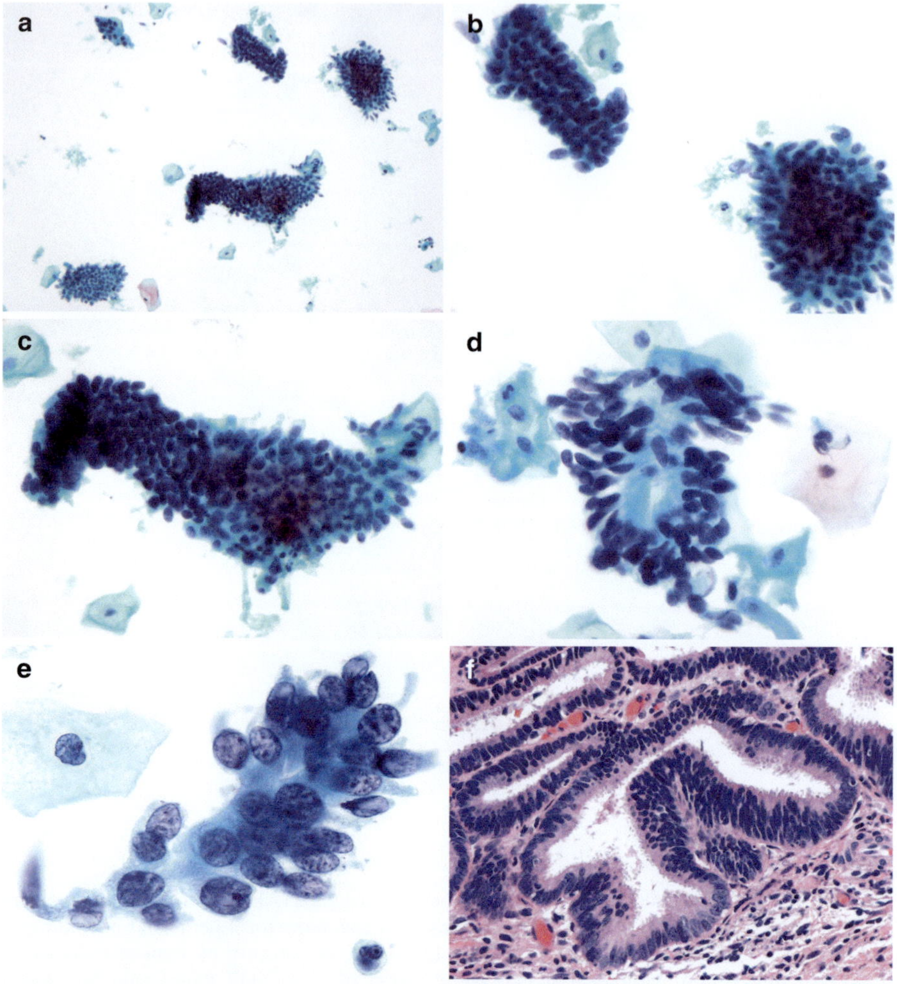

Fig. 2.19 Adenocarcinoma in situ (AIS). (**a**) Architecturally, AIS displays small and large HCGs; sheets, with nuclear pseudostratification; and flattened two-dimensional sheets with peripheral palisading. True "feathering" (nuclei at the periphery protruding beyond the confines of the cell due to extreme nuclear crowding and cohesion to basement membrane) is fairly specific for AIS and can be seen (TP). (**b**, **c**) Cytologically, nuclei of AIS are round to oval with notched and angular contours, irregularly thickened membranes, and hyperchromatic with coarsely stippled chromatin. Nucleoli are usually inconspicuous or small. Note nuclei crowding, feathering, and apoptotic bodies (**b** and **c**, TP). (**d, e**) Rosette and another feathered group show crisp nuclear changes of AIS (TP). (**f**) AIS on resection specimen shows complex branching glands lined by hyperchromatic, oval, and pseudostratified nuclei with eosinophilic cytoplasm (H&E). (**g**, **h**) AIS in SP displays small and large HCGs. Background is inflamed (SP). (**i**, **j**) Pseudostartified strips in two different cases (SP). False "feathering," or cytoplasmic tufts creating a feathered outline, is nonspecific and can be found in a variety of benign conditions such as tubal metaplasia

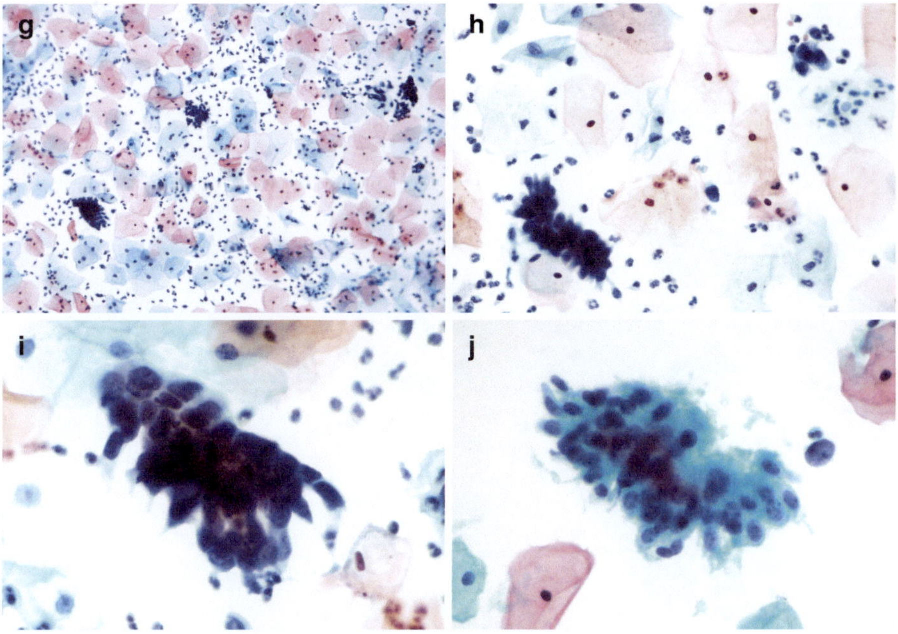

Fig. 2.19 (continued)

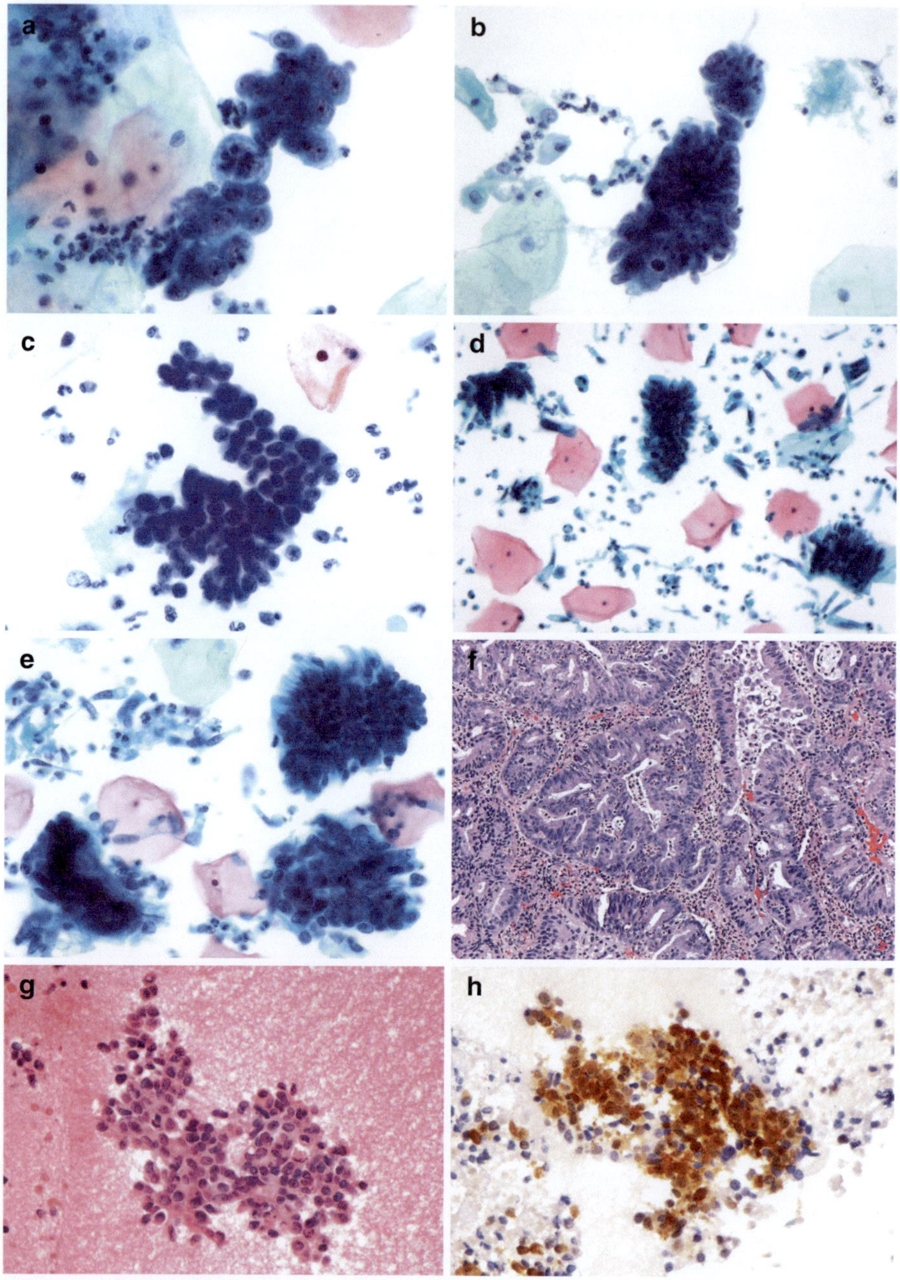
a
b
c
d
e
f
g
h

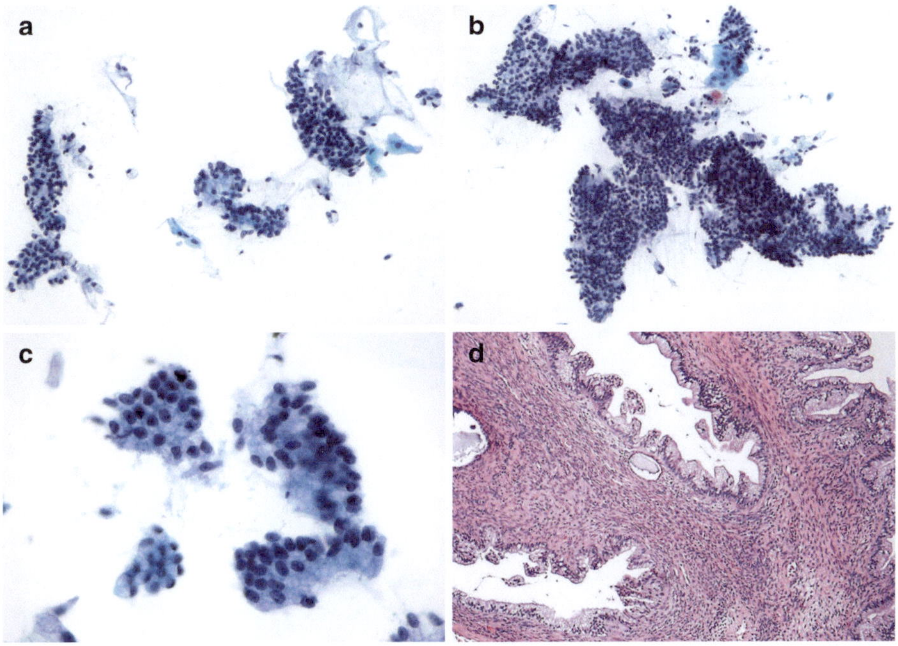

Fig. 2.21 Endocervical adenocarcinoma, minimal deviation type (MDA). (**a, b**) The endocervical cells show architectural variability and were dispersed as small and large two-dimensional groups and sheets with a honeycomb configuration and complex branching structures. Clumps of mucin are attached to some fragments (TP). (**c**) Honeycombing configuration of neoplastic endocervical cells with round to oval nuclei with subtle membrane irregularity, prominent nucleoli, and mucinous cytoplasm. Individual tumor cells are round to columnar with mucinous and vacuolated cytoplasm (TP). (**d**) Corresponding tissue section showing bland-appearing infiltrating glands with varying shapes (H&E). MDA is also known as adenoma malignum and is a rare tumor that causes diagnostic difficulty in both cytologic and surgical specimens due to its minimal cytologic atypia [24]

Fig. 2.20 Endocervical adenocarcinoma. (**a–c**) Usual type shows clusters and irregularly branching fragments of neoplastic cells with large round nuclei, stippled coarse chromatin, and macronucleoli, a key feature of invasion. Cytoplasm is finely vacuolated. Note the background inflammatory necrotic diathesis that "clings" to tumor cells (TP). TP is more sensitive in detecting endocervical and endometrial adenocarcinoma compared to conventional smears (72 vs. 41.5 %). (**d**) Endocervical adenocarcinoma in SP shows many single cells, HCGs with AIS-like features, except for the presence of many single neoplastic cells, diathesis, and prominent nucleoli. (**e**) Nuclei are similar to TP, cytoplasm appears dense, and nuclear to cytoplasmic ratio is high. Cells appear more elongated in SP (SP). (**f**) Endocervical adenocarcinoma on follow-up biopsy, showing crowded and disorganized glands (H&E). (**g, h**) In this case, a cell block was prepared from residual TP specimen vial which showed endocervical primary that showed strong diffuse positivity for p16 and monoclonal CEA immunostains (**g**, H&E and **h**, CEA)

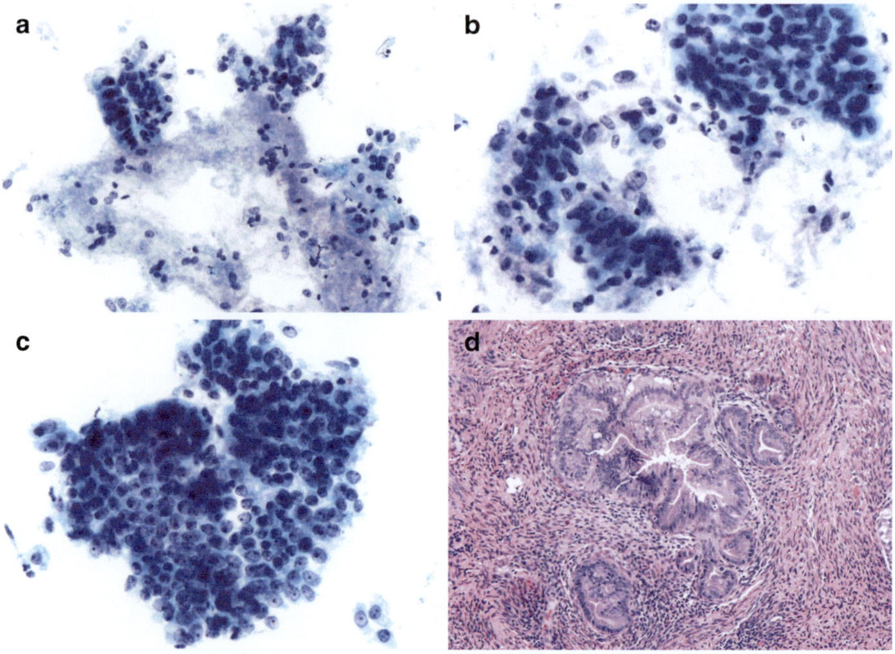

Fig. 2.22 Endocervical adenocarcinoma, gastric type. (**a**) The tumor cells show more architectural and cellular atypia compared to MDA. Cells are columnar, oval to round, and contain enlarged, round, or oval nuclei. Some peripheral palisading or nuclear "feathering" can be seen. Abundant mucin and necrotic debris are present in the background, attached to the neoplastic cells (TP). (**b**) The cytoplasm is pale bluish and either finely granular or vacuolated and foamy. The nuclei are crowded, are overlapping, and show clumped coarse chromatin and contain prominent nucleoli (TP). (**c**) Large honeycomb sheet shows highly atypical and crowded cells with prominent nucleoli (TP). (**d**) Infiltrating glands present in a haphazard arrangement (H&E). Unlike usual type of endocervical adenocarcinoma, this variant is typically negative for high-risk HPV [23]

a

b

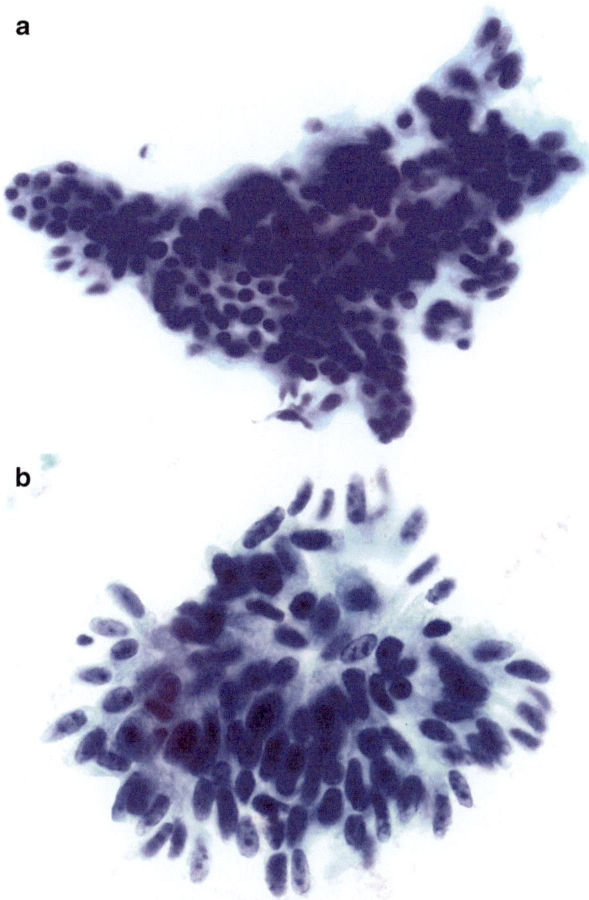

Fig. 2.23 Endocervical adenocarcinoma, villoglandular type. (**a**) Cells with high nuclear to cytoplasmic ratio and enlarged, hyperchromatic nuclei. The cells form a three-dimensional structure with irregular papillary-like projections (TP). (**b**) Prominent "feathering" can be seen around the entire fragment; the nuclei are elongated with distinct nucleoli. This is a papillary, well-differentiated adenocarcinoma with a good prognosis and cytologically bland features

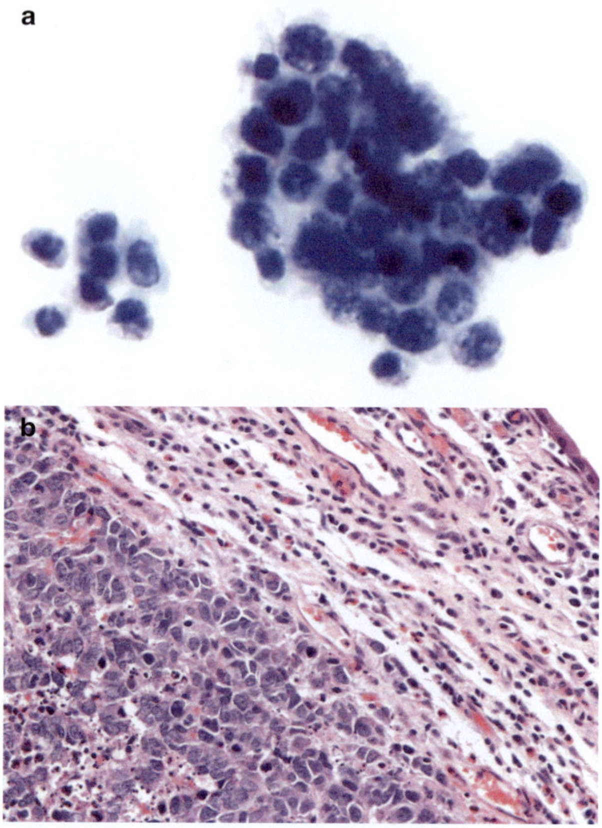

Fig. 2.24 Small cell carcinoma. (**a**) Small cells with minimal cytoplasm are closely clustered together such that the nuclei overlap or appear to "mold" into one another. The chromatin pattern has a "salt and pepper" appearance, suggesting the neuroendocrine nature of this tumor. Nucleoli are not evident. The background is clean, unlike on conventional smear which may contain prominent necrosis. Crush artifact or nuclear smearing was not seen (TP). (**b**) Small cells infiltrate the tissue (*bottom left*). Unlike the LBP, necrotic foci are readily identifiable (H&E). Small-cell carcinoma can be diagnosed on Pap tests and fine needle aspiration specimens on LBP. However, pathologists should be aware of certain morphologic alterations, particularly on TP, in order to avoid diagnostic pitfalls. These include less obvious or focal nuclear molding, more nuclear overlap; nucleoli may become visible or prominent; crush artifact or nuclear smearing may be absent or reduced. The latter may be represented by nuclear elongation with retained or smudged chromatin features

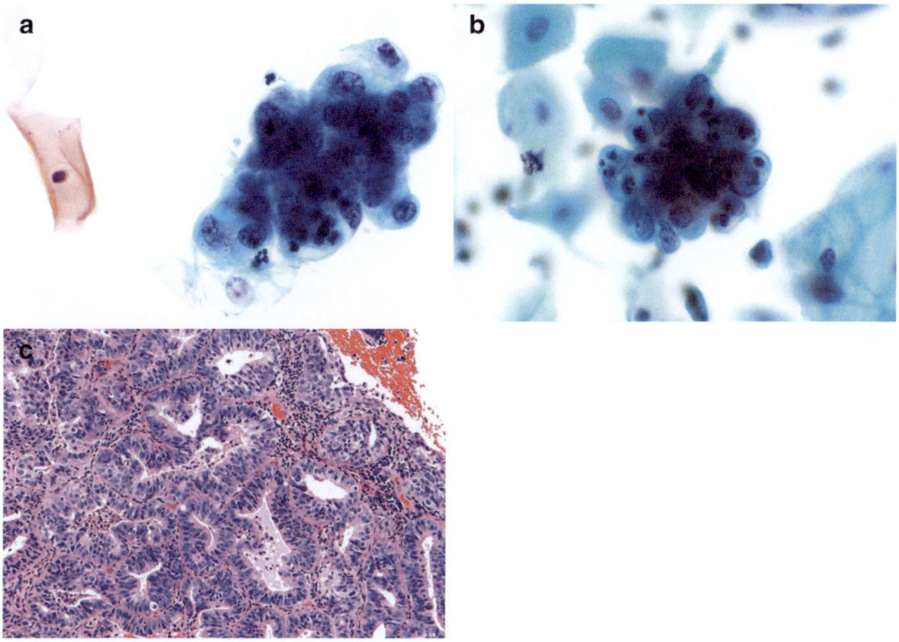

Fig. 2.25 Endometrial adenocarcinoma. (**a, b**) Three-dimensional cell groups with irregular nuclei that are large (more than 2× intermediate cell nuclei), angulated with open chromatin, nucleoli are visible. Cytoplasm is vacuolated and some cells show engulfed intact neutrophils. Tumor diathesis is not evident in either LBP (**a**, TP; **b**, SP). (**c**) Corresponding endometrial biopsy demonstrated a moderately differentiated endometrial adenocarcinoma with crowded, back-to-back glands with endometrioid morphology (H&E)

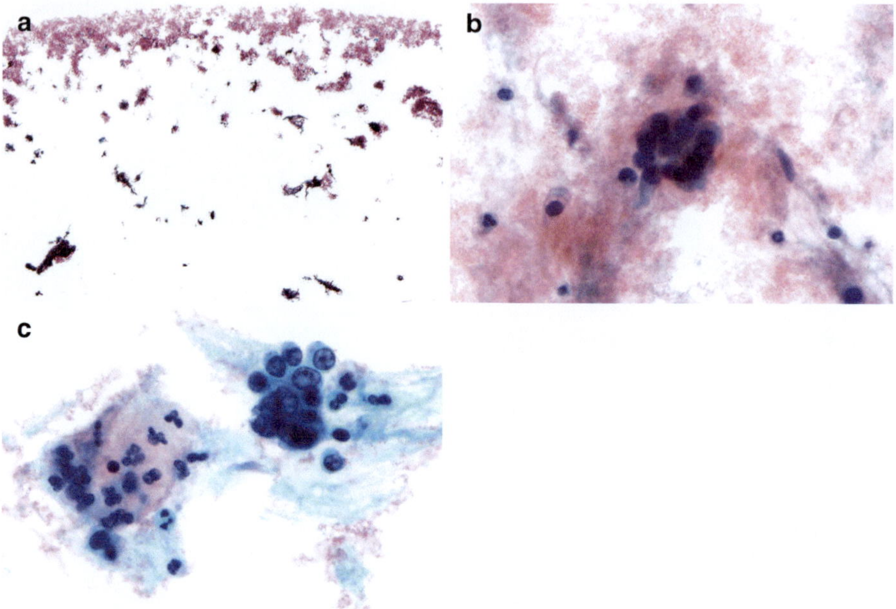

Fig. 2.26 Treatment with glacial acetic acid. (**a**, **b**) Excessive blood in a postmenopausal woman with bleeding has resulted in a "clog" of the TP filter. Note the blood forming a peripheral ring with the central portion of the slide showing rare HCGs trapped in blood and otherwise empty. In truth, a large amount of diagnostic material has not been transferred to the slide. (**c**) A repeat specimen following treatment with glacial acetic acid and CytoLyt clearly shows the endometrial adenocarcinoma. Unsatisfactory specimens due to blood should be reprocessed, especially when HCGs are present (**a**, TP; **b** and **c**, TP)

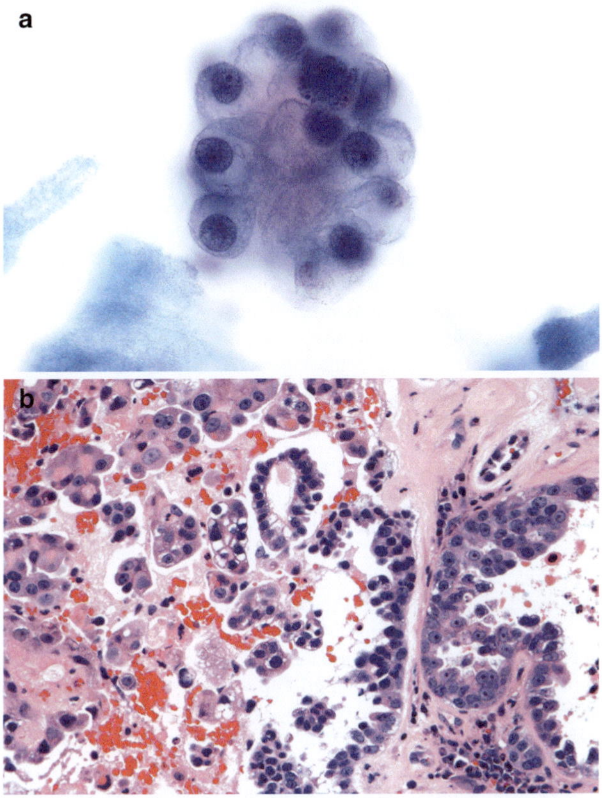

Fig. 2.27 Clear cell endometrial carcinoma. (**a**) Note the abundant cytoplasm, round nuclei with regular borders, and prominent nucleoli (TP). (**b**) The corresponding biopsy shows similar-appearing cells, with "hobnail" morphology and prominent nucleoli (H&E)

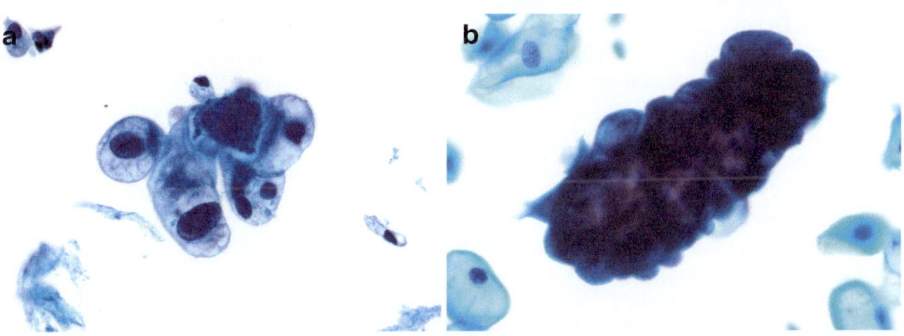

Fig. 2.28 Metastatic adenocarcinoma. (**a**) *Ovarian primary* shows malignant cells with abundant vacuolated cytoplasm and dark nuclei. Review of prior morphology and immunostains for WT1 and PAX-8, performed on cell block section, were confirmatory. (**b**) *Breast primary, ductal type*, shows a three-dimensional group of cells with enlarged malignant nuclei. Review of prior morphology and immunostain for GATA-3, performed on cell block section, were confirmatory. Background appears clean in both images (**a** and **b**, TP). Uterine metastasis of ductal and lobular carcinoma is rare. However, the latter has a propensity to metastasize to this site

Suggested Reading

1. Nayar R, Wilbur DC, editors. The Bethesda system for reporting cervical cytology: definitions, criteria, and explanatory notes. 3rd ed. Cham: Springer; 2015.
2. Hoda RS, Loukeris K, Abdul-Karim FW. Gynecologic cytology on conventional and liquid-based preparations: a comprehensive review of similarities and differences. Diagn Cytopathol. 2013;41:257–78.
3. Saslow D, Solomon D, Lawson HW, et al. ACS-ASCCP-ASCP cervical cancer guideline committee. American cancer society, American society for colposcopy and cervical pathology, and American society for clinical pathology screening guidelines for the prevention and early detection of cervical cancer. CA Cancer J Clin. 2012;62:147–72.
4. Wilbur DC, Henry MR. College of American pathologists. College of American pathologists practical guide to gynecologic cytopathology: morphology, management, and molecular methods. Northfield: College of American Pathologists; 2008. p. 255.
5. Arbyn M, Bergeron C, Klinkhamer P, Martin-Hirsch P, Siebers AG, Bulten J. Liquid compared with conventional cervical cytology: a systematic review and meta-analysis. Obstet Gynecol. 2008;111:167–77.
6. Hoda RS. Non-gynecologic cytology on liquid-based preparations: a morphologic review of facts and artifacts. Diagn Cytopathol. 2007;35:621–34.
7. Owens CL, Buist DS, Peterson D, Kamineni A, Weinmann S, Ross T, Williams AE, Stark A, Adams KF, Doubeni CA, Field TS. Follow-up and clinical significance of unsatisfactory liquid-based Papanicolaou tests. Cancer Cytopathol. 2015;123:59–65.
8. Haack LA, O'Brien D, Selvaggi SM. Protocol for the processing of bloody cervical specimens: glacial acetic acid and the ThinPrep Pap test. Diagn Cytopathol. 2006;34:210–3.
9. Aslan DL, McKeon DM, Stelow EB, Gulbahce HE, Kjeldahl K, Pambuccian SE. The diagnosis of trichomonas vaginalis in liquid-based Pap tests: morphological characteristics. Diagn Cytopathol. 2005;32:253–9.
10. Laudadio J, Hoda RS. Unique appearance of actinomyces on Thinprep Pap test. Diagn Cytopathol. 2006a;34:553–4.
11. Laudadio J, Hoda RS. Unique appearance of actinomyces on Thinprep Pap test. Diagn Cytopathol. 2006b;34:553–4.
12. Thrall MJ, Russell DK, Bonfiglio TA, Hoda RS. Use of the ThinPrep imaging system does not alter the frequency of interpreting Papanicolaou tests as atypical squamous cells of undetermined significance. Cytojournal. 2008;5:10.
13. Rao R, Molina D, Halligan AM, Vakil B, Alperstein SA, Hoda RS. Negative computer-imaged ThinPrep Pap test and positive hybrid capture2 HPV co-testing results: a quality assurance review. Diagn Cytopathol. 2015;43:763–9.
14. Gilani SM, Tashjian R, Fathallah L. Cervical cytology with a diagnosis of atypical squamous cells, cannot exclude high-grade squamous intraepithelial lesion (ASC-H): a follow-up study with corresponding histology and significance of predicting dysplasia by human papillomavirus (HPV) DNA testing. Arch Gynecol Obstet. 2014;289:645–8.
15. Renshaw AA, Mody DR, Wang E, Haja J, Colgan TJ. Hyperchromatic crowded groups in cervical cytology-differing appearances and interpretations in conventional and ThinPrep preparations: a study from the College of American pathologists interlaboratory comparison program in cervicovaginal cytology. Arch Pathol Lab Med. 2006;130:332–6.
16. Clark SB, Dawson AE. Invasive squamous-cell carcinoma in ThinPrep specimens: diagnostic clues in the cellular pattern. Diagn Cytopathol. 2002;26:1–4.
17. Bai H, Sung CJ, Steinhoff MM. ThinPrep Pap test promotes detection of glandular lesions of the endocervix. Diagn Cytopathol. 2000;23:19–22.
18. Keyhani-Rofagha S, Vesey-Shecket M. Diagnostic value, feasibility, and validity of preparing cell blocks from fluid-based gynecologic cytology specimens. Cancer. 2002;96:204–9.
19. Wood MD, Horst JA, Bibbo M. Weeding atypical glandular cell look-alikes from the true atypical lesions in liquid-based Pap tests: a review. Diagn Cytopathol. 2007;35:12–7.

20. Ozkan F, Ramzy I, Mody DR. Glandular lesions of the cervix on thin-layer Pap tests. Validity of cytologic criteria used in identifying significant lesions. Acta Cytol. 2004;48:372–9.
21. Chivukula M, Austin RM, Shidham VB. Evaluation and significance of hyperchromatic crowded groups (HCG) in liquid-based paps. Cytojournal. 2007;4:2.
22. Schorge JO, Hossein Saboorian M, Hynan L, Ashfaq R. ThinPrep detection of cervical and endometrial adenocarcinoma: a retrospective cohort study. Cancer. 2002;96:338–43.
23. Fulmer CG, Hoda RS, Pirog EC, Park KJ, Holcomb K. Cytomorphology of gastric type cervical adenocarcinoma on a ThinPrep Pap test: report of a p16-Positive tumor case. Diagn Cytopathol. 2016;44:710–3.
24. Hissong E, Yoxtheimer LM, Pacecca A, Hoda RS. Cytology of minimal deviation endocervical adenocarcinoma (adenoma malignum) on a ThinPrep Pap test. Diagn Cytopathol. 2016;44:552–5.
25. Wagner DG, Weisensel J, Mentrikoski MJ, Leo SD, Bonfiglio TA, Hoda RS. ThinPrep Pap test of endocervical adenocarcinoma with lymph node metastasis: report of a case in a 17-year-old woman. Diagn Cytopathol. 2010;38:633–8.
26. Norimatsu Y, Sakamoto S, Ohsaki H, Ozaki S, Yokoyama T, Shimizu K, Yanoh K, Akiyama M, Bamba M, Kobayashi TK. Cytologic features of the endometrial adenocarcinoma: comparison of ThinPrep and BD SurePath preparations. Diagn Cytopathol. 2013;41:673–81.

Urinary Tract Cytology

3

Introduction

In 2016, approximately 76,000 new cases of bladder carcinoma will be diagnosed in the United States. Of these, around 58,000 will be in men and 18,000 in women. Approximately 16,000 bladder cancer-related deaths will occur in 2016. About 50 % of bladder cancer cases are diagnosed while it is noninvasive, and around 35 % are diagnosed when it has invaded into deeper layers of the bladder, but the disease is still confined to the organ (American Cancer Society, Cancer Facts and Figures, 2016).

More than 90 % of the bladder cancers are urothelial in origin. The examination of urine cytology is an efficient method to diagnose primary and recurrent carcinoma of urinary tract, i.e., bladder, urethra, ureters, and renal pelves. A urinary sample can be collected as voided specimen, through urinary catheter, or by cystoscope using instrumental lavage (washings) or brushings.

Urine cytology specimens comprise the largest proportion of non-gynecological specimens in most cytology laboratories and are increasingly being processed as liquid-based preparations (LBP). In this chapter, cytology and differential diagnosis of nonneoplastic and neoplastic lesions of the lower urinary tract collected via different methods are illustrated. The utility of ancillary tests such as immunocytochemistry is also addressed.

The Paris System (TPS) for Reporting Urinary Cytology

The Paris System (TPS) is a standardized international reporting system for urinary cytology specimens. The system was conceived by participants of two urine cytology symposia in 2013 at the 18th International Congress of Cytology (ICC) held in Paris. TPS guidelines are based on consensus, global participation, and evidence-based practice. The primary emphasis of TPS is the detection of

R.S. Hoda et al., *Diagnostic Liquid-Based Cytology*,
DOI 10.1007/978-3-662-53905-7_3, © Springer-Verlag GmbH Germany 2017

high-grade urothelial carcinoma (HGUC), with the understanding that urinary cytology has low sensitivity, specificity, and reproducibility for the detection of low-grade urothelial neoplasms (LGUN). The reporting system is sponsored by the American Society of Cytopathology (ASC) and the International Academy of Cytology (IAC). TPS diagnostic categories include *negative for HGUC* (NHGUC); *atypical urothelial cells* (AUC); *suspicious for HGUC* (SHGUC), *HGUC*, *LGUN*; and *other malignancies, both primary and secondary*. Many of these categories utilize strict cytomorphological criteria and in particular focus on nuclear-to-cytoplasmic ratio, hyperchromasia, nuclear border irregularity, and chromatin quality (Figs. 3.1–3.24).

- *NHGUC*: The negative category implies that no cytomorphological features of HGUC are identified in the specimen. If a specimen does not meet the criteria of AUC, milder atypical changes can be classified into this category. NHGUC includes cases in which LGUN cannot be excluded. If there is a cause for atypia without the presence of any features of HGUC (i.e., urolithiasis, treatment-related changes, viral cytopathic, or other reactive changes), this category is indicated.
- *AUC*: In order to be classified as AUC, the cells should be non-degenerated, non-superficial cells with a nuclear-to-cytoplasmic ratio of at least 0.5 and at least one of the following: mild to moderate hyperchromasia, nuclear border irregularities, and/or irregular clumpy chromatin. The number of AUC is not a criterion, but evidence suggests the number correlates with risk.
- *SHGUC*: This category is used for more severe atypia that quantitatively falls short of a definitive diagnosis of HGUC – 5–10 HGUC cells. The cells should be non-degenerated, non-superficial cells with a nuclear-to-cytoplasmic ratio of at least 0.5 and at least moderate hyperchromasia. In addition, the suspicious cells should have either markedly irregular nuclear borders or irregular clumpy chromatin.
- *HGUC*: These cells may be from carcinoma in situ (CIS) or papillary HGUC. The requirements are nuclear-to-cytoplasmic ratio of 0.7 or above, moderate-to-severe hyperchromasia, irregular nuclear borders, and coarse, clumpy chromatin.
- *LGUN*: A combined cytological term for low-grade papillary urothelial neoplasms include urothelial papilloma, papillary urothelial neoplasm of low malignant potential (PUNLMP), low-grade papillary urothelial carcinoma (LGUC), and flat, low-grade intraurothelial neoplasia. This diagnosis technically falls under the category of NHGUC, but may be used if clinical and cystoscopy findings correlate with a papillary lesion and no features of HGUC are identified.
- *Other malignancies, both primary and secondary*: Used for non-urothelial malignancies, both primary and secondary – non-urothelial carcinoma including squamous cell, adeno-, and small cell carcinoma and non-epithelial malignancies including lymphoma, melanoma, sarcoma, and metastatic tumors to lower urinary tract.

Indication, Collection, and Laboratory Processing of Cytological Samples

The principal indications for the use of cytology in disorders of urinary tract (bladder, urethra, ureters, and renal pelves) are as follows:

1. The diagnosis of HGUC: The cytological techniques are of limited value in the identification of LGUC.
2. The routine follow-up and monitoring of patients with a history of bladder cancer, for the early detection of recurrences. The current standard of care consists of cystoscopy and urine cytology evaluation every 3–6 months for the first year and at reduced intervals subsequently.

Methods of Specimen Collection

The principal methods of specimen collection are:

- Voided urine
- Catheterized urine (CU)
- Direct sampling techniques
  - Bladder washings, barbotage, or lavage
  - Collection by retrograde catheterization of ureters
  - Direct brushings of ureters and renal pelves

The method of specimen collection and processing depends on the clinical circumstances and goal of the examination.

Voided Urine

This is the easiest method of cytological investigation of the urinary tract. A small volume of the collected specimen is preserved in liquid-based preservative medium, or if received fresh, it can be refrigerated for 24 h until processing.

Catheterized Urine

The specimens are collected via a urinary catheter and preserved and processed as described for voided urine specimens.

Direct Sampling Techniques

Bladder Washings (Barbotage or Lavage)

This technique may be utilized during cystoscopic examination to obtain well-preserved cells from high-risk patients. It is the specimen of choice for DNA ploidy analysis. Bladder washings have significantly better diagnostic yields.

Retrograde Catheterization of Ureters or Renal Pelves

This procedure is used to establish the nature of a radiologically detected space-occupying lesion of ureter or renal pelvis, including a calculus, blood clot, neoplasms, or inflammatory masses. In the ureters, other causes include a stricture or extraneous pressure.

The Direct Brushing Procedure

Brushing is performed through a ureteral catheter to investigate space-occupying lesions in the ureters or renal pelves.

Ileal Conduit Urine

After cystectomy for HGUC, an artificial bladder is often constructed from a segment of the small intestine, usually the ileum. The status of these organs must be monitored after treatment due to the propensity of urothelial tumors to recur.

Laboratory Processing of Urinary Specimens

Several techniques are available for processing urinary specimens for microscopy including LBP, cytocentrifugation, and membrane filtration. The two LBP, ThinPrep (TP, Hologic, Bedford, MA) and SurePath (SP, BD Diagnostics, Burlington, NC), are illustrated in Chap. 1.

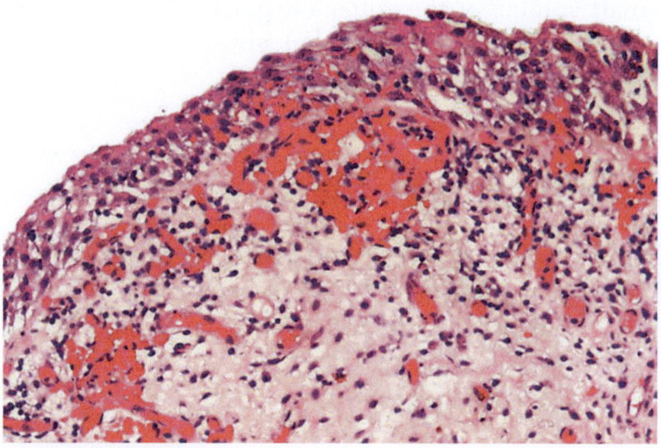

Fig. 3.1 Normal urothelial histology. Normal urothelial lining in distended bladder. There are approximately seven layers of urothelial cells. The surface is lined by larger superficial cells with abundant cytoplasm ("umbrella cells"). Capillary vessels are present immediately beneath the epithelium in the lamina propria (H&E)

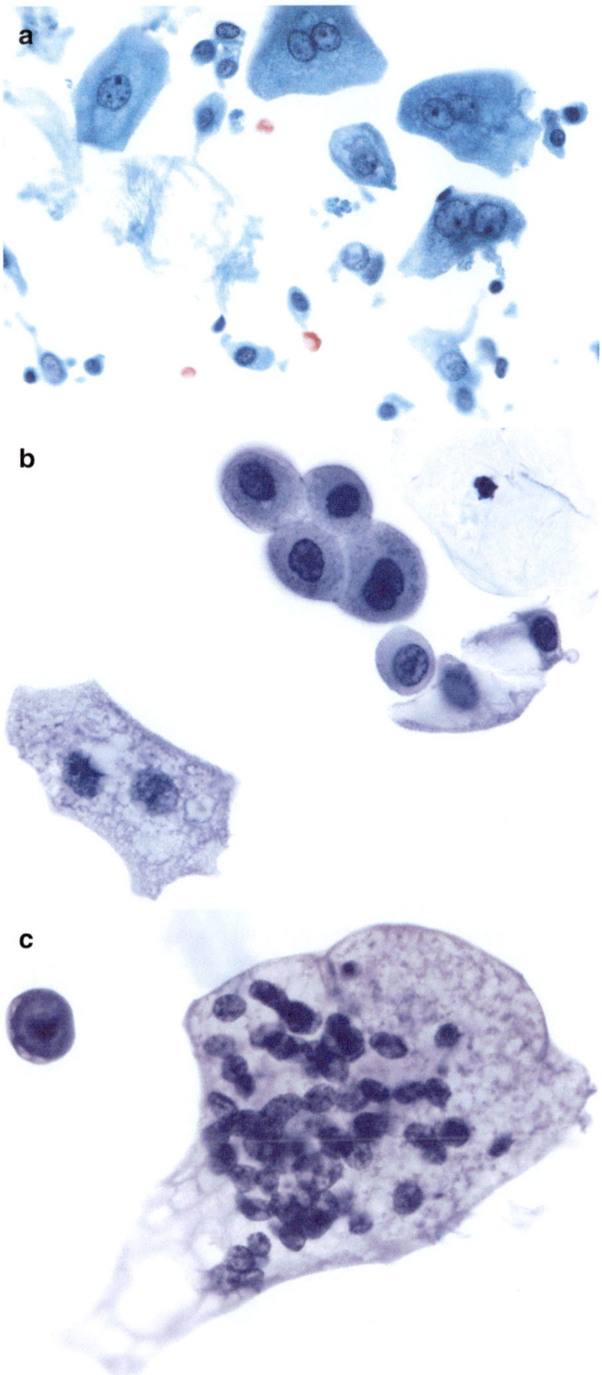

Fig. 3.2 Normal urothelial cells in voided urine. (**a**) *Superficial (umbrella) cells*: normal voided urine may contain relatively few urothelial cells. These cells are of various types, shapes, and sizes and comprise superficial (umbrella) cells and a few cells from the intermediate and deeper urothelial cell layers. Umbrella cells are the predominant cell type in voided urine and are 20–30 μm in diameter, with one flat plane. They are somewhat similar in size to cervicovaginal intermediate squamous cells, but may be larger. Umbrella cells may be mono-, bi-, or multinucleated. The nuclei are central and round with regular finely granular chromatin and occasional small nucleoli (SP). (**b**) *Intermediate cells originating from the deeper layers of the urothelium:* resemble parabasal squamous cells from lower genital tract in size and configuration. These cells are often spherical or round with scanty basophilic cytoplasm and spherical bland nucleus. They desquamate singly or in clusters (TP). (**c**) Multinucleated umbrella cells are rare in voided urine, except after diagnostic and therapeutic procedures. They are normally seen in specimens from ureter or renal pelves. Giant umbrella cells may be mistaken as neoplastic (TP)

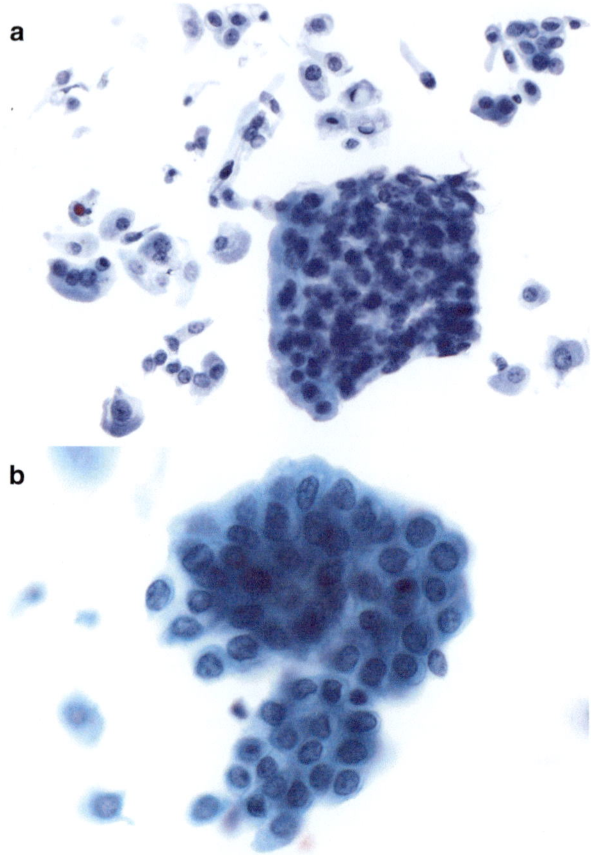

Fig. 3.3 Catheterized urine. (**a**) The cytological features of catheterized urine (CU) are usually similar to voided samples, except for higher cellularity, more sheets, clusters and urothelial tissue fragments, and background inflammation. Some umbrella cells show reactive features with prominent single and multiple nucleoli. The nuclei however are regular and pale (TP). Catheterized urine is considered instrumented specimen and therefore may have artifacts not seen in voided urine, particularly the presence of increased urothelial tissue fragments which may be at first worrisome for low-grade urothelial neoplasia (LGUN). Kannan and Bose [1] found that evenness of fragment borders and the presence of cytoplasmic collars were distinguishing features of instrument artifact. (**b**) Since urothelial tissue fragments have some three-dimensionality, inflammatory cells and other background cells may appear on a different plane of focus in SP. This does not occur in TP, as fragments seen in those preparations are flat (SP)

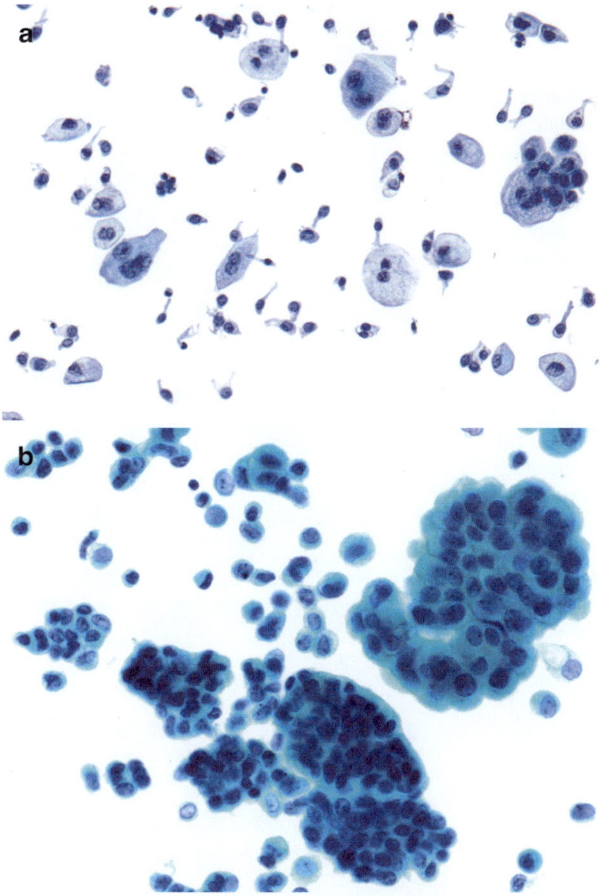

Fig. 3.4 Washing specimens in the absence of disease. (**a**) Bladder washing specimens are usually highly cellular and often include a larger number of urothelial tissue fragments and single urothelial cells. In fact, a specimen that is sparsely cellular may indicate an inadequate sample. Prather et al. [2] determined that a cutoff of 20 well-visualized, well-preserved urothelial cells per ten high-power fields increased the positive predictive value of bladder wash (TP). (**b**) Deeper urothelial cells are more frequent and may form tightly packed three-dimensional clusters in a "papillary" configuration with relatively smooth borders. In these instances, the cells are often more elongated with regular, oval nuclei, bland chromatin, and small nucleoli. Umbrella cells usually outline these groups. These large clusters are commonly mistaken as representing a papillary tumor, particularly, if the cytologist is unaware of how the specimen was procured (SP). Washings and brushings from upper urinary tract are richly cellular and generally appear similar to bladder wash. Large sheets of urothelial cells may be forcibly removed from surface of ureters or renal pelves resulting in complete denudation of ureteral surface

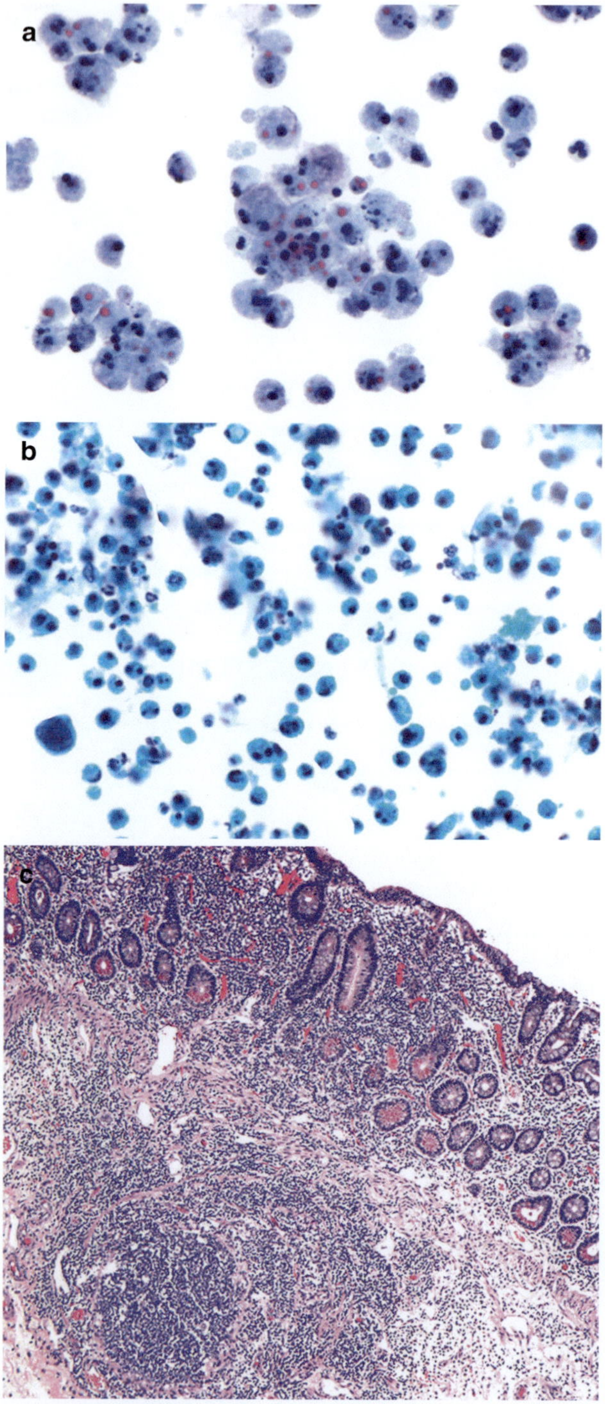

Fig. 3.5 Ileal conduit urine. (**a**) Ileal conduit urine (ICU) used in monitoring patients after cystectomy for bladder cancer normally contains a rich population of poorly preserved intestinal cells. The intestinal cells may have vacuolated or granular degenerative cytoplasm and fragmented and/or pyknotic nuclei. Background mucin and Paneth cells may sometimes be seen. Wolinska and Melamed [3] studied ICU and were able to trace the transition from intestinal epithelial cells in clusters to degenerating rounded cells that resemble histiocytes (TP). In less degenerated samples, well-preserved columnar mucus-producing cells with peripheral and round dark nuclei are noted. (**b**) In this specimen, numerous degenerative cells distract from rare malignant cells in the field; this patient had a recurrence of HGUC (SP). (**c**) In histological specimens taken from a benign ileal conduit, progressive shortening and ultimately flattening of villi can be seen, as well as an increased number of goblet cells (H&E)

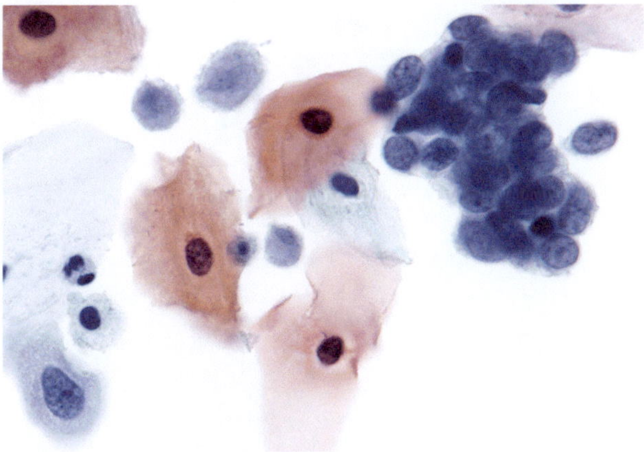

Fig. 3.6 Squamous cells. Urine samples are often contaminated by squamous cells from anogenital region, trigone, or urethra, though this contamination may be limited by procuring a sample through catheterization or instrumentation. Superficial squamous cells are characterized by a small, pyknotic nucleus and abundant cytoplasm, and intermediate cells have a slightly larger, vesicular nucleus. Smaller parabasal cells may be seen, which have an increased nuclear-to-cytoplasmic ratio and share some cytomorphological overlap with urothelial cells. Note cluster of deeper urothelial cells. Pear-shaped *Trichomonas vaginalis* organisms indicate contamination from the lower gynecologic tract. Occasionally atypical squamous cells (ASC) may be seen, such as koilocytes; in these instances, they should be noted and clinical correlation should be suggested. It can be difficult to isolate the origin of ASC, as they could represent any lesion from condyloma to HGUC with squamous differentiation in voided urine (see Fig. 3.23a, b) (TP). Owens and Ali [4] followed 32 patients with ASC in the urine; 8 patients had bladder carcinoma with squamous differentiation, 2 had cervical carcinoma, and 22 (69 %) had benign follow-up

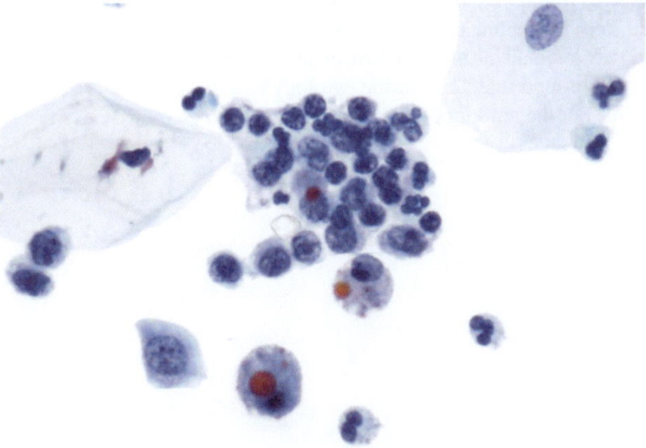

Fig. 3.7 Acute and chronic cystitis. Mixed inflammatory cells comprise of polymorphonuclear leukocytes, lymphocytes, and eosinophils. Desquamated benign urothelial cells can be seen. These findings are usually nonspecific and require clinical correlation. If necrotic debris is observed, it should raise awareness for tumor in which only necrotic cells are present. In some instances, repeated infections may cause cystitis follicularis, which may result in the presence of dendritic cells with prominent nuclear chromatin and prominent nucleoli, lymphoid aggregates, reactive urothelial cell atypia, and glandular cells, indicating glandular metaplasia [5]. Melamed-Wolinska bodies (dense, round eosinophilic inclusions) can be observed in degenerating urothelial cells. Multiple inclusions of varying sizes may be present and are thought to represent degenerative lysosomal aggregates (TP)

Fig. 3.8 Crystals. Rarely, true uric acid crystals can have a variety of shapes and sizes and are typically yellow or colorless. Possible shapes include rhomboids, hexagonal plates, needles, and rosettes. They typically form under acidic conditions and are not necessarily a sign of an underlying pathological process (TP). Calcium carbonate crystals, which usually appear as large yellow-brown or colorless spheroids with radial striations, are also commonly seen in urinary specimens. They can also be seen as smaller crystals with round, ovoid, or dumbbell shapes. Triple phosphate crystals (struvite) appear as "coffin lids" and form in alkaline urine. They are classically associated with *Proteus mirabilis* infection. For the most part, crystals are simply distracting noncellular element seen in urine specimens. Wright and Halford [6] found that TP of urine cytology had the same numbers of casts and debris as cytospin preparations, but the amount of crystals and red blood cells (RBCs) were reduced in TP

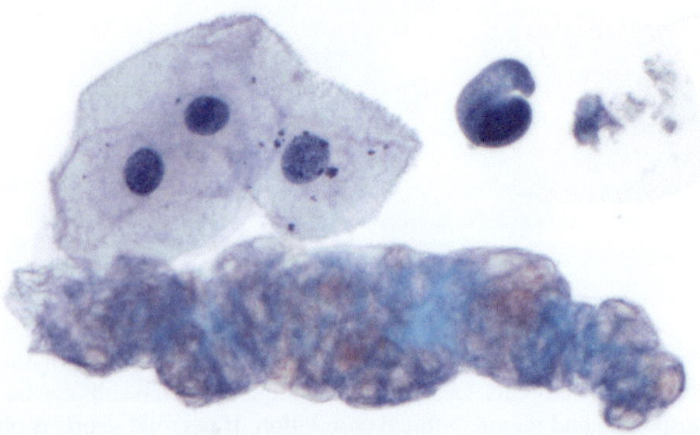

Fig. 3.9 Renal casts. Casts arise from the kidney. In this particular example, the compact cast is made of RBCs (TP). Other commonly seen casts include granular casts, white blood cell (WBC) casts, bacterial casts, epithelial casts (consisting of renal tubular cells), and cellular casts (consisting of a mixture of cells). Casts are formed under various conditions affecting the kidney, and there is no evidence that their presence is associated with HGUC

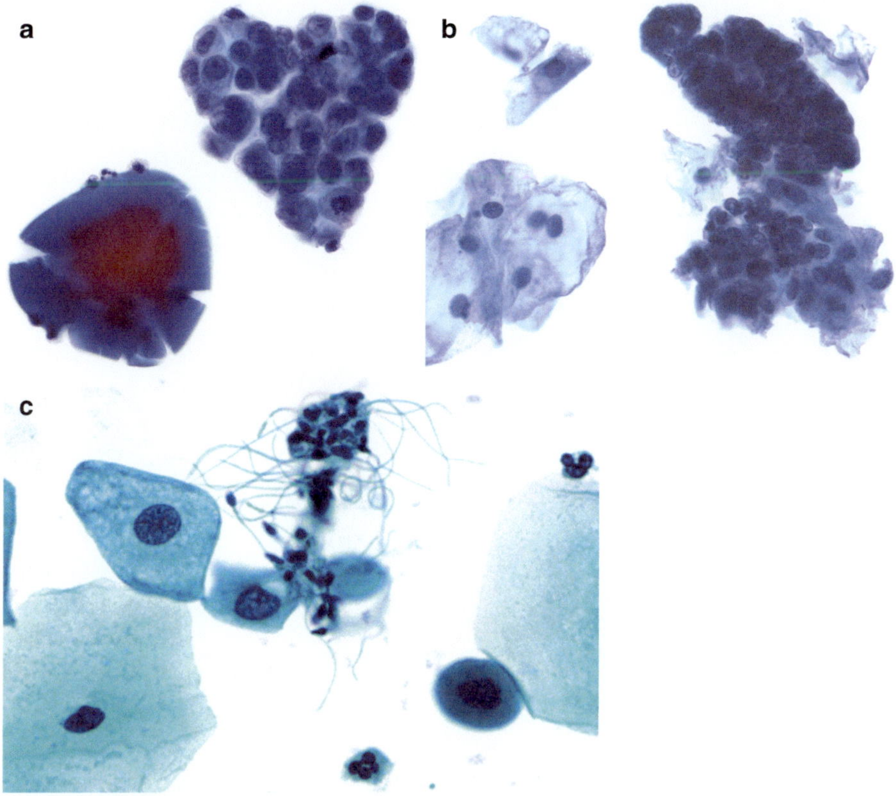

Fig. 3.10 Contaminants and distractors. (**a**) Corpora amylacea appears as cellular, smooth, oval, lamellated basophilic structures that have cracks at the periphery due to calcification. They are of no clinical significance (TP). (**b**) Benign endometrial cells (EMC) may be seen in voided urine in women as part of vaginal contamination. Correlation with menstrual history would be prudent. EMC may also be present in women with endometriosis involving urinary bladder (TP). Bohlmeyer and Shroyer [7] found that the cytologic appearance of EMC from bladder washings may not necessarily be similar to those seen in cervical smears but may be similar. EMC that have been recently washed from the wall of the bladder will be seen on the slide as three-dimensional groups of cells with a high nuclear-to-cytoplasmic ratio. They found that uniformity of EMC is the most useful cytological feature on which to base a benign diagnosis, (**c**) sperm may be present in urine secondary to retrograde ejaculation. In some instances, they may be engulfed by macrophages (spermiophages) (SP)

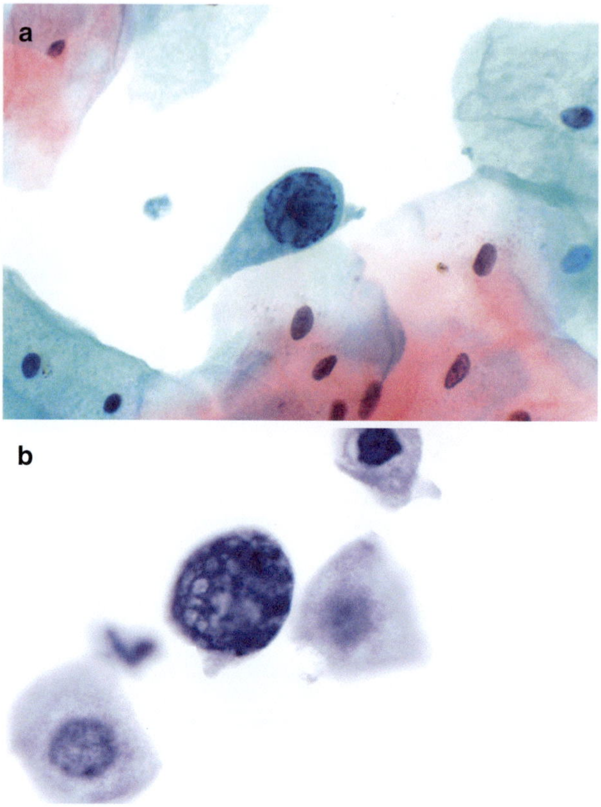

Fig. 3.11 Polyomavirus (BK). (**a**) "Decoy cells" are characterized by large cells with a single, spherical, regular, and hyperchromatic nucleus containing large homogeneous basophilic intranuclear inclusion that occupies the entire nucleus leaving only a narrow rim of chromatin. The name "decoy" refers to the cytomorphological similarity these cells have to HGUC (e.g., enlarged hyperchromatic nucleus and high nuclear-to-cytoplasmic ratio). "Decoy" cells are usually round but may also be elongated with a cytoplasmic tail "comet" cells (SP). (**b**) The inclusions may also dissolve, presumably because the viral particles leach out, leaving behind a peculiar network of coarse nuclear chromatin that is as diagnostic of infection. Note the large round cells with enlarged round nucleus and scant rim of cytoplasm (TP). As the infection persists, the inclusions become pale and less basophilic and acquire a homogeneous appearance. Confirmation of the diagnosis can be performed by urine polymerase chain reaction (PCR) analysis or a positive immune reaction with an anti-SV40 monoclonal antibody. However, the morphologic manifestations of this viral infection are often characteristic so that it is rarely necessary to resort to immunochemistry for diagnosis. If no atypical changes worrisome for HGUC are present, the specimen should be diagnosed as *negative for HGUC* (NHGUC). In instances where one cannot distinguish between the two, a diagnosis of at least *atypical urothelial cells* (AUC) is appropriate. Allison et al. [8] demonstrated an increased rate of HGUC on follow-up among specimens containing cells with morphology of BK infection, suggesting that morphological overlap with HGUC can lead to underdiagnosis

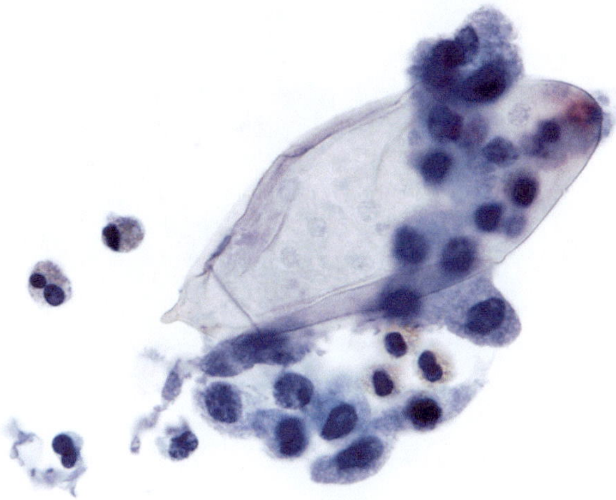

Fig. 3.12 *Schistosoma haematobium*. The ovum is oval with a thick transparent capsule and shows a sword-shaped protrusion located at the narrow end of the ovum, known as the terminal spine (TP). *S. haematobium* has a known association with primary bladder squamous cell carcinoma, though its presence in urine is often associated with granular debris and inflammation. In some specimens, maturing ciliated larval forms (miracidium) may be seen ([9]). Primary squamous carcinoma of the bladder may also be associated with other factors

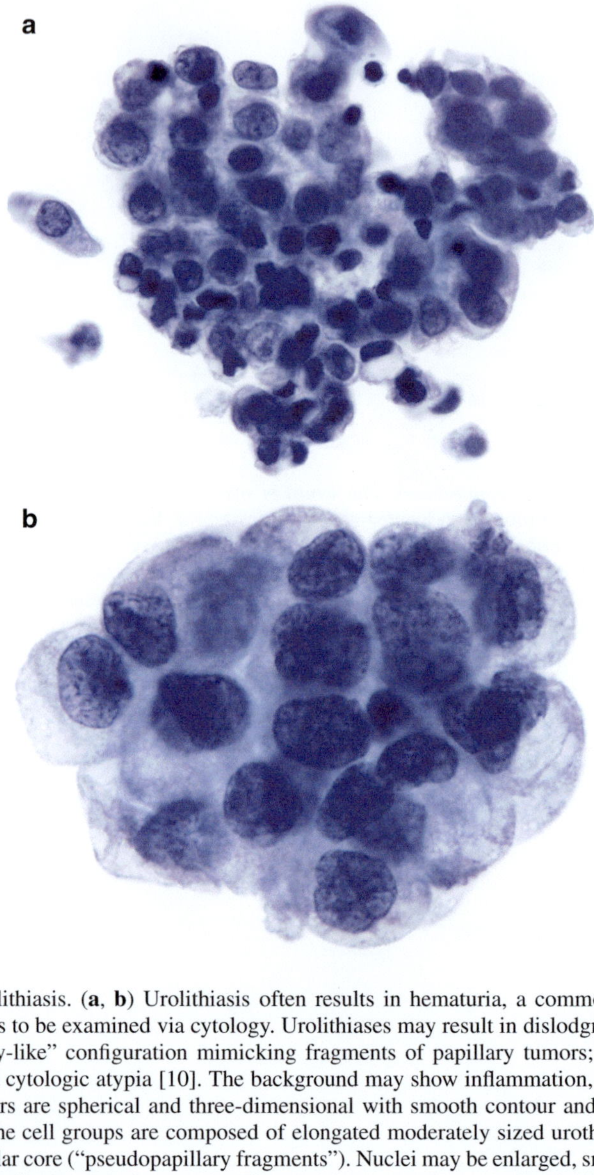

Fig. 3.13 Urolithiasis. (**a**, **b**) Urolithiasis often results in hematuria, a common indication for urine specimens to be examined via cytology. Urolithiases may result in dislodgment of UTF that show "papillary-like" configuration mimicking fragments of papillary tumors; some fragments may show mild cytologic atypia [10]. The background may show inflammation, RBCs, and crystals. The clusters are spherical and three-dimensional with smooth contour and peripheral cytoplasmic rim. The cell groups are composed of elongated moderately sized urothelial cells which lack fibrovascular core ("pseudopapillary fragments"). Nuclei may be enlarged, smooth, and round with nucleoli indicating reactive atypia. The cytoplasm may be vacuolated (TP). Occasionally, crystals may be embedded within cell clusters. Patients with calculi may also have concurrent urothelial carcinoma

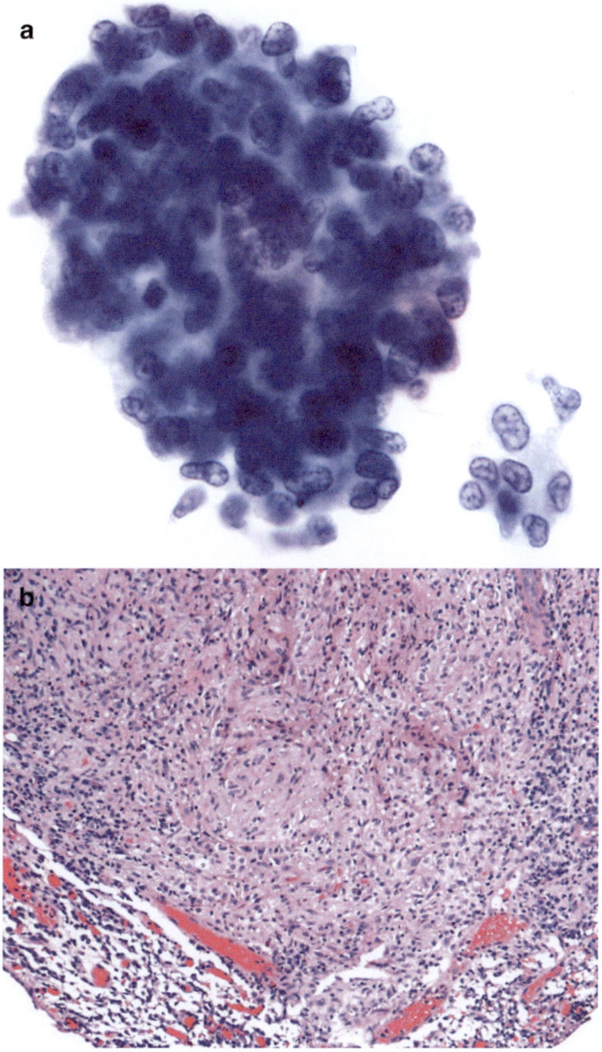

Fig. 3.14 Bacillus Calmette-Guérin (BCG) treatment effect. (**a**) A well-formed noncaseating granuloma, appearing as a tightly clustered group of histiocytes and lymphoid cells. While the three-dimensional quality of this group may be worrisome, the kidney-shaped histiocyte nuclei are reassuring (TP). Patients with early noninvasive urothelial carcinoma are often treated by intravesical BCG. As a result, subsequent pathology may demonstrate noncaseating granulomas and inflammatory cells. Takashi et al. [11] found that treatment changes included translucent nuclei, prominent nucleoli, and cytoplasmic vacuolization with eosinophilic inclusions and that these changes could be distinguished from HGUC. (**b**) The corresponding bladder biopsy shows noncaseating granulomas with abundant eosinophilic cytoplasm. The urothelial lining is attenuated (H&E)

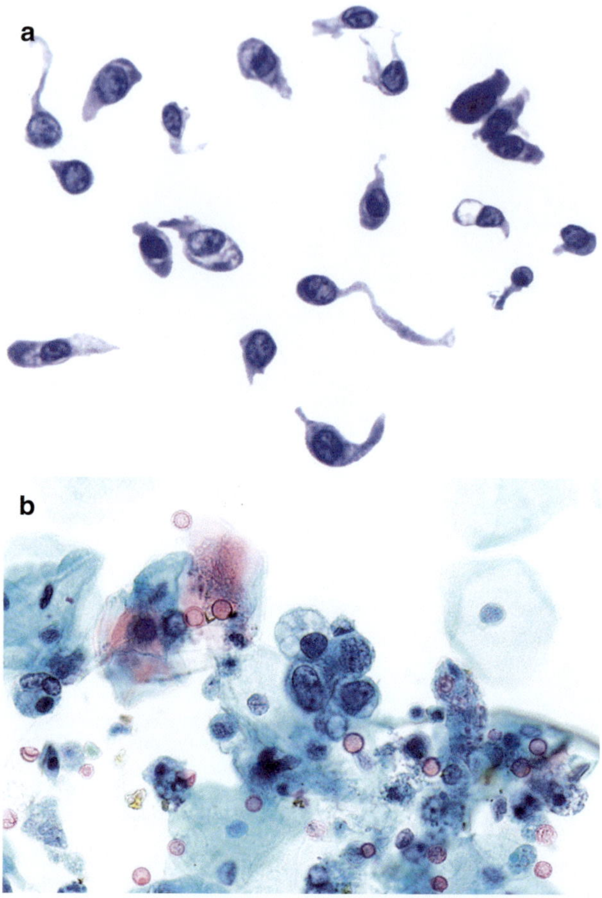

Fig. 3.15 Atypical urothelial cells (AUC). (**a**) Ureteral washings with dispersed elongated urothelial cells, some with a "cercariform" appearance, i.e., cytoplasmic processes which have a nontapering, flattened, bulbous, or fishtail-like end and nucleated globular body. Occasional cells have a nuclear-to-cytoplasmic ratio that approaches 0.7, but only mild hyperchromasia is present. Furthermore, nuclei are relatively regular. A follow-up tissue biopsy was diagnosed as LGUC (TP). (**b**) The AUC lies in a background of degenerative changes. The nuclear-to-cytoplasmic ratio is above 0.7 and there are mild nuclear border irregularities. Follow-up biopsy was diagnosed as HGUC (SP). Although TPS indicates a nuclear-to-cytoplasmic ratio of 0.5, in the examples shown here, the ratio was around 0.7. The reason, they are termed AUC is because of lack of other criteria for suspicious or malignant diagnosis. It is important to accurately estimate nuclear-to-cytoplasmic ratio along with other nuclear features. A recent study indicates that pathologists tend to overestimate nuclear-to-cytoplasmic ratio [12]

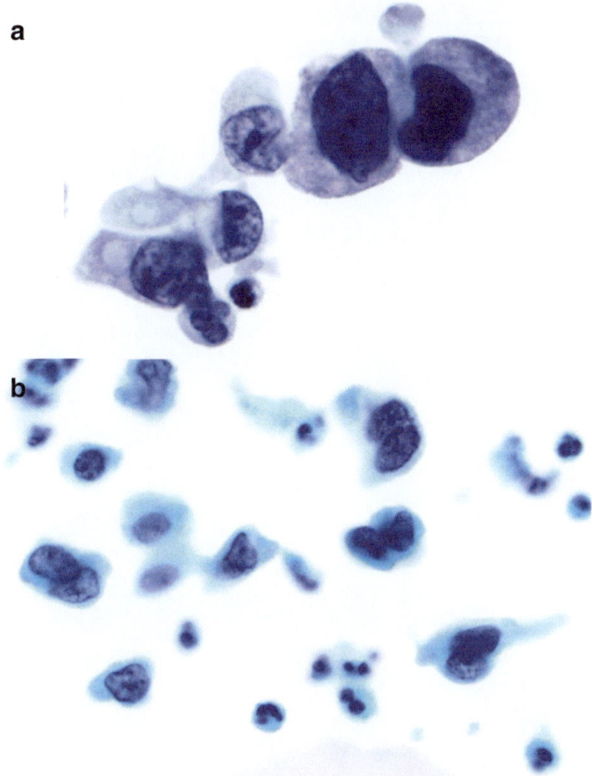

Fig. 3.16 Suspicious for high-grade urothelial carcinoma (SHGUC). (**a**) TPS requires SHGUC cells to have a nuclear-to-cytoplasmic ratio of at least 0.5, with irregular nuclear borders, hyperchromasia, and coarse chromatin. This category is used when 5–10 cells meet the criteria of HGUC. This group of urothelial cells has variable nuclear-to-cytoplasmic ratios, but several of them are over 0.5. Some cells have marked hyperchromasia (TP). (**b**) Scattered urothelial cells which are only slightly enlarged (compared to adjacent neutrophils). However, the cells have irregular nuclear borders, have clumpy chromatin, and have nuclear-to-cytoplasmic ratios which in some cases approach 0.7. The cells are not as hyperchromatic as seen in (**a**) (SP). The SHGUC category and its equivalents have been shown to predict a high rate of subsequent HGUC, though this may not be proven until much later [13]

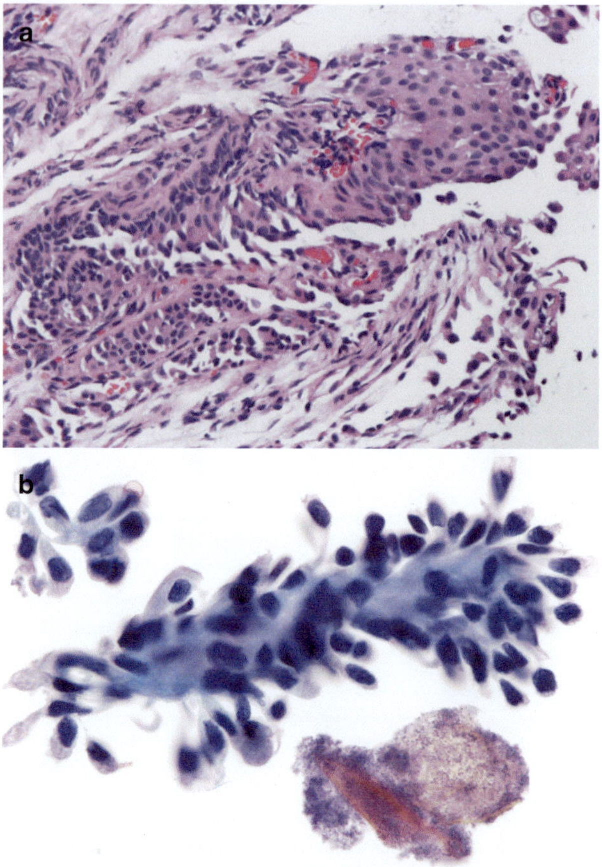

Fig. 3.17 Low-grade urothelial carcinoma (LGUC). (**a**) Histology of a typical LGUC with thin branches carrying capillary vessels (H&E). (**b**) This voided urine specimen is from a patient with LGUC. This tissue fragment appears to contain a fibrovascular core with attached monotonous urothelial cells. The cells are elongated, giving them a cercariform appearance. While the cells appear hyperchromatic, they have mild nuclear border irregularities and granular chromatin. In such an instance, TPS category LGUN can be used, but only if high-grade features are not seen in any other field TP. (**c**) Cellular specimen demonstrating small cells with high nuclear-to-cytoplasmic ratios but round and regular nuclear borders and bland chromatin pattern. Note the presence of cercariform cells (SP). LGUN includes urothelial papilloma, papillary urothelial tumor of low malignant potential (PUNLMP), and LGUC. Urine cytology has been shown to have poor sensitivity, specificity, and reproducibility for LGUN, as the cytological features are nonspecific and often overlap with benign and reactive conditions [14]. Several studies have reported cytological features purported to be associated with LGUC in LBP. Xin et al. [15] found high nuclear-to-cytoplasmic ratio, irregular nuclear borders, and homogenous cytoplasm to be the best predictors of LGUC on TP. According to TPS, the only reliable feature for a definitive diagnosis of LGUN in urine cytology is the presence of a fibrovascular core and the absence of high-grade features. Furthermore, such lesions are allowed to be classified as NHGUC. (**d**) LGUC with hypercellular, three-dimensional fragment containing fibrovascular cores and a complex papillary structure. Note the low nuclear-to-cytoplasmic ratios and bland chromatin of the neoplastic cells (SP)

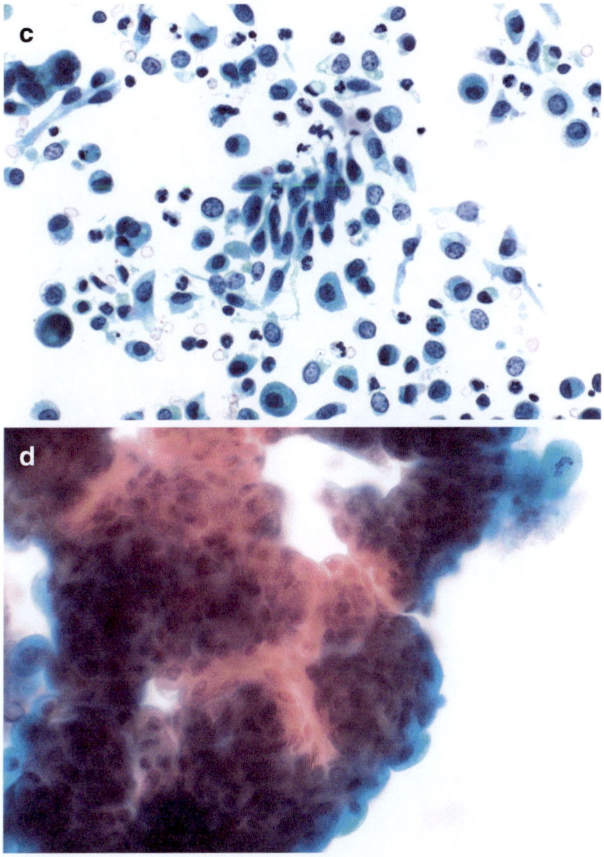

Fig. 3.17 (continued)

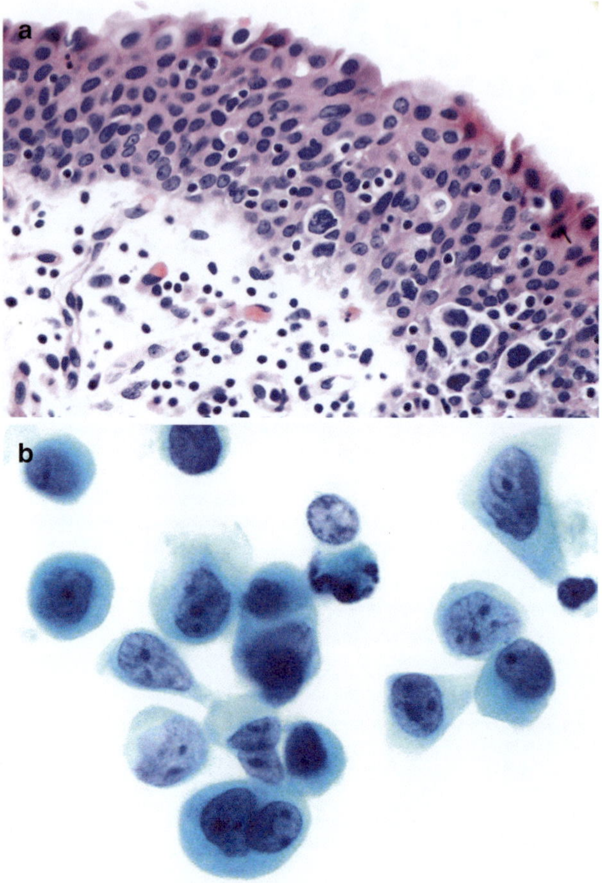

Fig. 3.18 Carcinoma in situ (CIS). (**a**) Histology of CIS shows nuclear border irregularities and anisonucleosis in some cells. In many instances, a tissue biopsy may show denuded urothelial lining, which by itself can be worrisome for UCC (H&E). (**b**) This field demonstrates a group of urothelial cells, some of which meet the criteria for HGUC. Most cells have a clumpy chromatin pattern with nuclear border irregularities. Some cells have marked hyperchromasia, and only a few meet the nuclear-to-cytoplasmic ratio cutoff of 0.7. This is not unusual, as the majority of carcinoma cells in HGUC will not meet the latter TPS criteria due to pleomorphism seen in HGUC. It is important to note that cytology cannot distinguish between CIS and papillary HGUC and cannot determine whether HGUC or CIS has become invasive. Therefore, the use of the HGUC category suggests the presence of either a papillary invasive and noninvasive HGUC or CIS (SP). Although not used as TPS criteria for HGUC, other features described include the presence of prominent nucleoli, isolated malignant cells, and extensive necrosis [16]

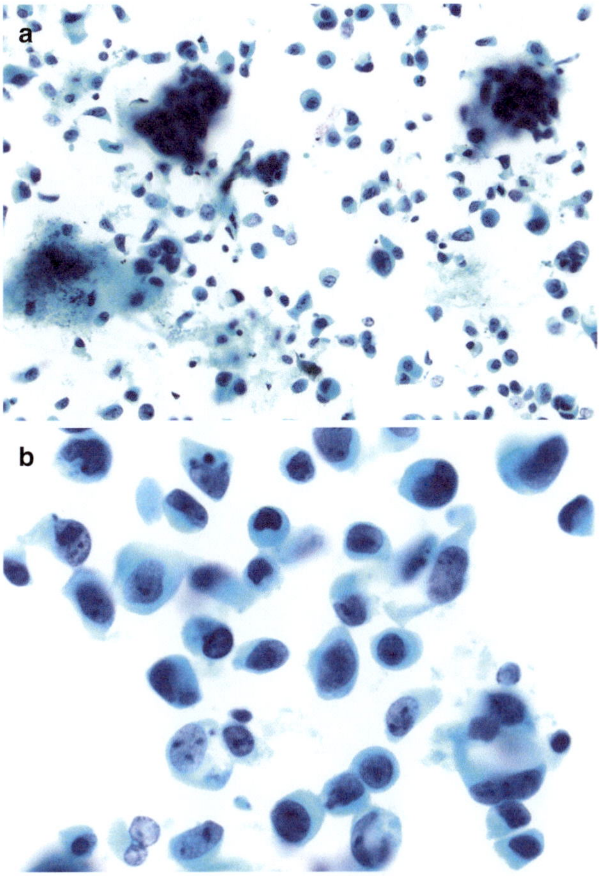

Fig. 3.19 High-grade urothelial carcinoma (HGUC). (**a**) The cellularity of this specimen may be appropriate for bladder washing, but the cells have dramatic pleomorphism even at low magnification. Closer examination reveals a "cell-in-cell" pattern in which malignant cells appear to engulf other cells. This is commonly seen in HGUC, though it is not a criterion used by TPS. It may be secondary to entosis, in which one cell engulfs another cell. Several hyperchromatic tissue fragments are seen which appear atypical, though this is difficult to assess, especially in SP preparations in which three-dimensional fragments may obscure features of individual cells (SP). (**b**) The examination of fields containing individual atypical cells (rather than fragments) at higher magnification is important in determining whether high-grade features are present. In this case, the cells have nuclear border irregularities, coarse chromatin, and greatly variable nuclear-to-cytoplasmic ratios. While there is no doubt that these are HGUC cells, note that few cells have nuclear-to-cytoplasmic ratio at or above 0.7 (SP)

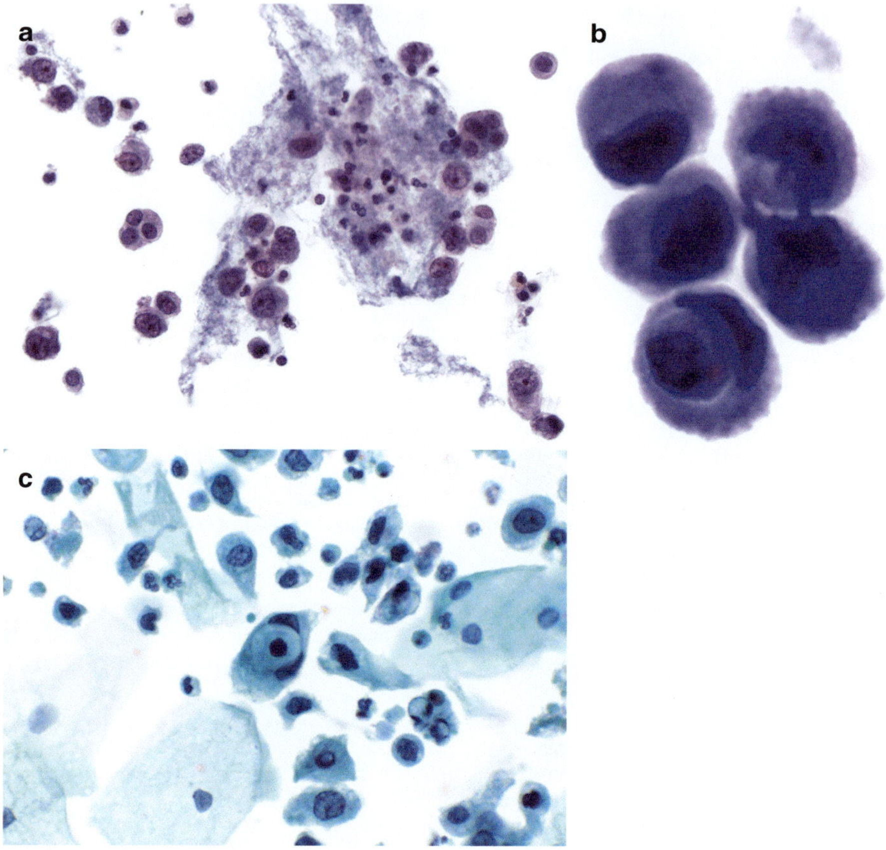

Fig. 3.20 HGUC. (**a**) Voided urine from invasive papillary HGUC shows malignant singly dispersed cells and small clusters. The cells are pleomorphic, and only rare cells reach nuclear-to-cytoplasmic ratio of 0.7. Most malignant cells are hyperchromatic with coarse chromatin, but some have regular nuclear borders. The background shows marked inflammation and necrosis that clings to tumor cells described as "clinging" diathesis in TP. In this case, some malignant cells are embedded in diathesis (TP). (**b**) The high-powered view shows malignant cells from invasive HGUC. While the nuclei are dark and opaque ("India ink nuclei"), their angulated "winged" shape results in nuclear-to-cytoplasmic ratios below 0.7. One pair of cells shows "cell-in-cell" phenomenon (TP). (**c**) While not a criterion of TPS, "pearl formation" seen here is often seen in HGUC (SP). (**d**) The subsequent biopsy showed papillary HGUC (H&E)

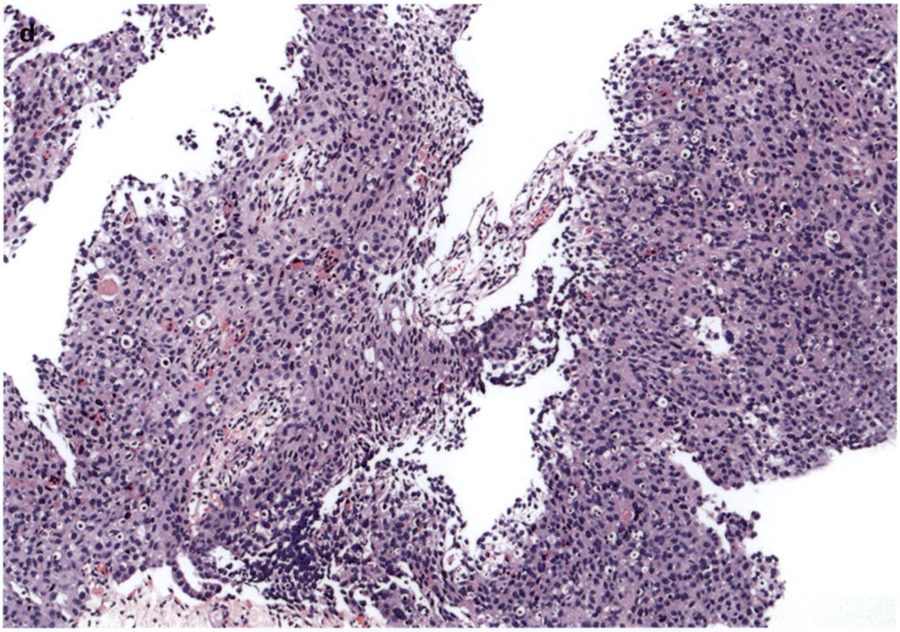

Fig. 3.20 (continued)

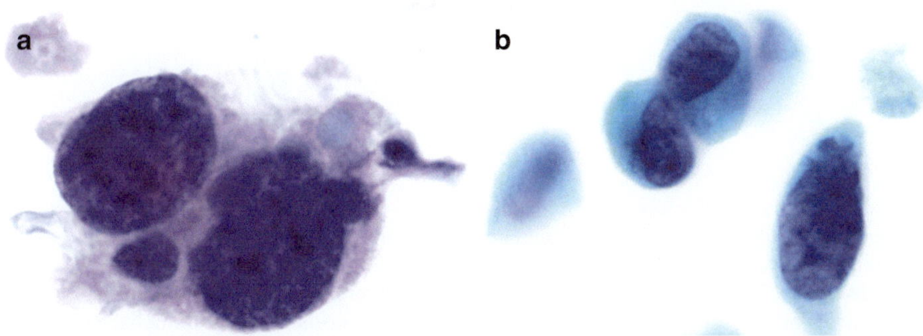

Fig. 3.21 High-grade urothelial carcinoma. (**a**) In TP preparation, malignant cells are large; nuclei show coarse, clumpy chromatin, multiple nucleoli, an irregular membrane, and an elevated nuclear-to-cytoplasmic ratio (TP). (**b**) In SP preparation, malignant cells are large with vesicular nuclei, thick irregular nuclear membranes, small nucleoli, and elevated nuclear-to-cytoplasmic ratios. Because SP preparation has more three-dimensionality than TP, large malignant cells often have an expanded Z-axis; when nuclei are in focus, smaller background cells become out of focus (SP)

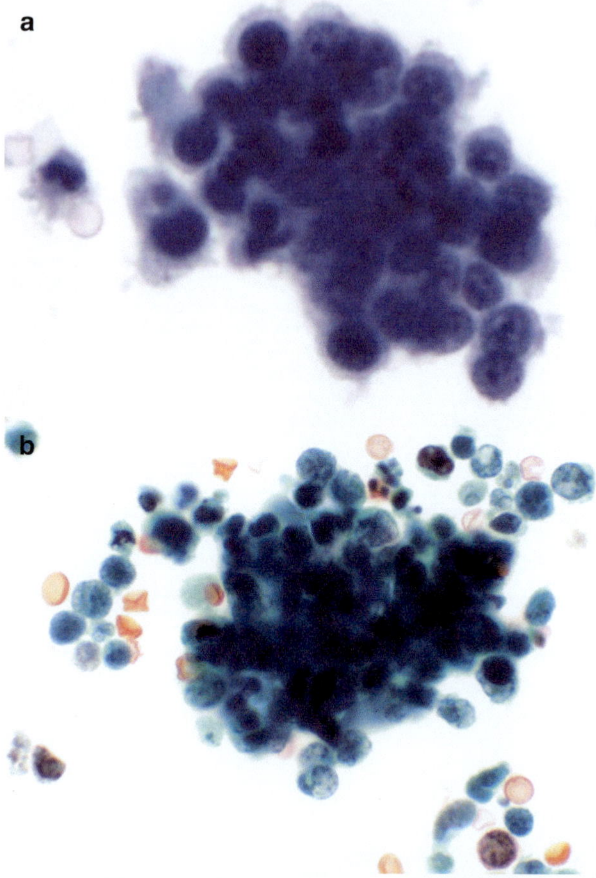

Fig. 3.22 Small cell carcinoma of the bladder. (**a**) This voided urine shows a three-dimensional cluster of small round *blue* cells which are larger than neutrophils. The cells may have discernible or barely visible cytoplasm. The nuclei show subtle molding and appear regular, hyperchromatic with coarse and clumped chromatin and nucleoli. Rare apoptotic cells are present that appear darker and smaller. Necrosis is not evident. Differential diagnosis includes urothelial carcinoma, lymphoma, and small-cell carcinoma. These features are not typically seen in urothelial carcinoma. Lymphoma cells are more dyscohesive and mostly single. Neuroendocrine marker for synaptophysin and TTF-1, performed on additional TP preparations, were positive, and a diagnosis of small cell carcinoma was rendered (TP). In LBP, many of the morphological clues of small cell carcinoma, as seen in direct smears, are attenuated. Background necrosis may be clumped. Crush artifact may only be seen as nuclear elongation. If the malignant cells are mostly dyscohesive, they may blend into the background and be at first mistaken for inflammatory cells at low magnification. However, higher magnification will reveal cells to be larger than benign lymphocytes [17]. (**b**) This group of small cell carcinoma cells is loosely cohesive, the cells have neuroendocrine-type chromatin, nucleoli are not visible, and apoptotic cells are scattered within the group. At the edges of the fragment, spherical *blue* granular debris represents necrotic tumor cells which still cling to viable cells (SP). HGUC can also be present with a small cell carcinoma component

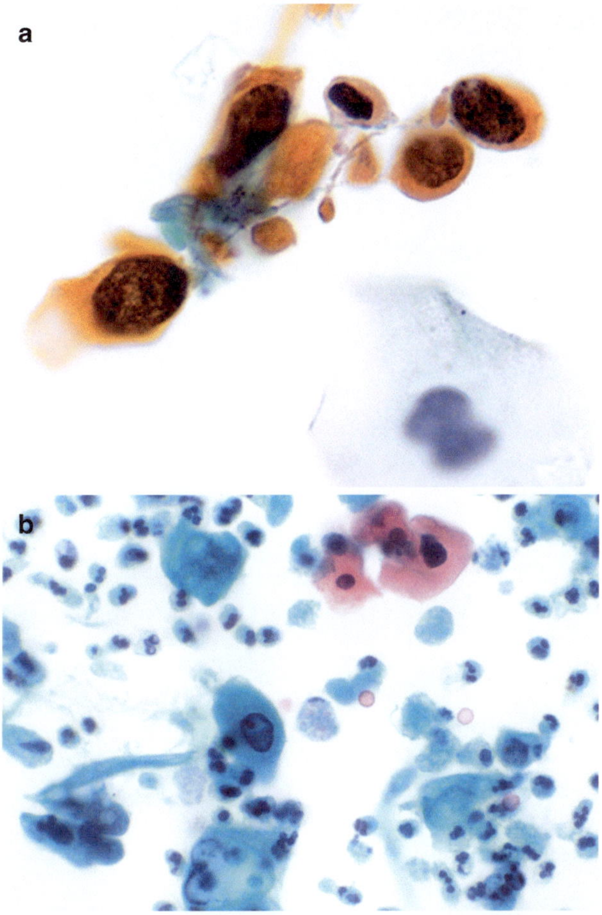

Fig. 3.23 Non-urothelial bladder tumors. (**a**, **b**) *Squamous cell carcinoma of the bladder*: Malignant cells with squamous differentiation can be seen here, as evidenced by the orangeophilic cytoplasm, dark pyknotic or opaque nuclei with irregular borders, and dense rigid cytoplasm. In SP, the specimen comprises of mostly tumor cells, some of which are necrotic (**a**, TP; **b**, SP). Squamous cell carcinoma may be primary in the bladder or even extend into urinary tract from lower female genital tract or penile primary. HGUC may also be present. Even though only malignant squamous cells may be seen, an underlying HGUC component may not be present in cytological specimen or even the biopsy. Diagnosis of pure squamous cell carcinoma is usually rendered on resection. (**c**, **d**) *Adenocarcinoma of the bladder*. (**c**) This voided urine specimen contains a cluster of carcinoma cells with soft granular cytoplasm and hyperchromatic rounded nuclei with prominent nucleoli. "Clinging" diathesis is prominent. Patient had a urethral primary (TP). Single prominent nucleoli are rarely seen in HGUC, and this finding should cause one to also consider metastatic (prostate or renal cell) carcinoma. (**d**) A cluster of cells resembles adenocarcinoma seen anywhere else. Three-dimensional groups of large cells with high nuclear-to-cytoplasmic ratios, prominent nucleoli, and marked anisonucleosis (TP). In this case, the patient had a urachal adenocarcinoma. Bardales et al. [18] found that the cytology of primary bladder adenocarcinoma was less characteristic than that of secondary adenocarcinoma

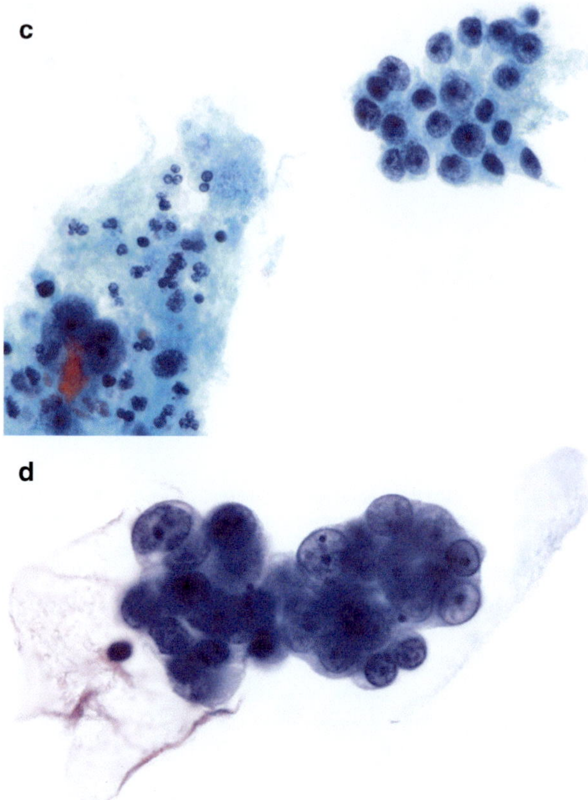

Fig. 3.23 (continued)

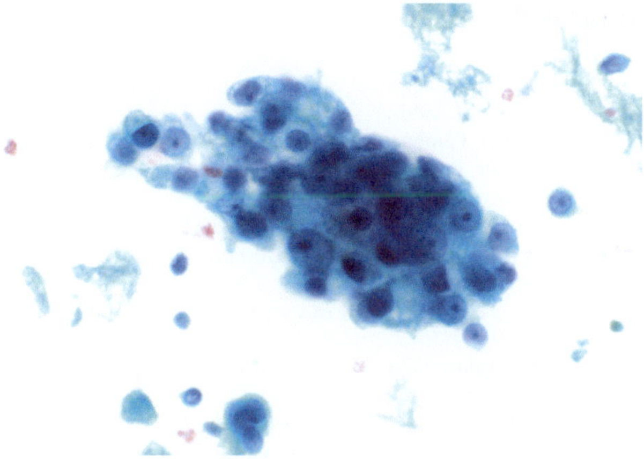

Fig. 3.24 Metastatic malignancies to the urinary tract. Prostatic adenocarcinoma involving urinary tract may be seen in urine specimens as it can gain access to the urinary tract through direct invasion. Typically, patients have a history of prostate carcinoma. The classic appearance of prostate carcinoma includes malignant cells in acinar formations with a round nucleus and prominent nucleolus and foamy cytoplasm without greatly increased nuclear-to-cytoplasmic ratio. Prominent nucleoli may be seen in renal cell carcinomas or melanoma, but is not a typical feature of HGUC. Bardales et al. [18] have reported that prostate cancer in urine may form syncytial and acinar arrangements. Immunostudies (e.g., PSA) can be used to confirm the site of origin. In some instances, prostate carcinomas may include, or consist entirely, of a small cell component (SP). As with other metastases to urinary tract, a history of primary tumor has usually been established. TPS categorization of malignancies is under the "other" categories, and if a primary HGUC cannot be excluded, "poorly differentiated carcinoma of uncertain origin" can be diagnosed

Suggested Reading

1. Kannan V, Bose S. Low grade transitional cell carcinoma and instrument artifact. A challenge in urinary cytology. Acta Cytol. 1993;37:899–902.
2. Prather J, Arville B, Chatt G, Pambuccian SE, Wojcik EM, Quek ML, Barkan GA. Evidence-based adequacy criteria for urinary bladder barbotage cytology. J Am Soc Cytopathol. 2015;4:57–62.
3. Wolinska WH, Melamed MR. Urinary conduit cytology. Cancer. 1973;32:1000–6.
4. Owens CL, Ali SZ. Atypical squamous cells in exfoliative urinary cytology: clinicopathologic correlates. Diagn Cytopathol. 2005;33:394–8.
5. Zaharopoulos P. Cytologic manifestations of cystitis follicularis in urine specimens. Diagn Cytopathol. 2002;27:205–9.
6. Wright RG, Halford JA. Evaluation of thin-layer methods in urine cytology. Cytopathology. 2001;12:306–13.
7. Bohlmeyer TJ, Shroyer KR. Endometriosis of the bladder: cytologic findings and differentiation from transitional cell carcinoma. Acta Cytol. 1996;40:383–4.
8. Allison DB, Olston MT, Lilo M, Zhang ML, Rosenthal DL, VandenBussche CJ. Should the BK polyomavirus cytopathic effect be best classified as atypical or benign in urine cytology specimens? Cancer Cytopathol. 2016;124:436–42.
9. Waugh MS, Perfect JR, Dash RC. *Schistosoma haematobium* in urine: morphology with ThinPrep method. Diagn Cytopathol. 2007a;35:649–50.
10. Onur I, Rosenthal DL, VandenBussche CJ. Atypical urothelial tissue fragments in noninstrumented voided urine specimens are associated with low but significantly higher rates of urothelial neoplasia than benign-appearing urothelial tissue fragments. Cancer Cytopathol. 2015;123:186–92.
11. Takashi M, Schenck U, Koshikawa T, Nakashima N, Ohshimad S. Cytological changes induced by intravesical Bacillus Calmette-Guérin. Urol Int. 2000;64:74–81.
12. Vaickus LJ, Tambouret RH. Young investigator challenge: the accuracy of the nuclear-to-cytoplasmic ratio estimation among trained morphologists. Cancer Cytopathol. 2015;123:524–30.
13. VandenBussche CJ, Sathiyamoorthy S, Owens CL, Burroughs FH, Rosenthal DL, Guan H. The Johns Hopkins Hospital template for urologic cytology samples: parts II and III: improving the predictability of indeterminate results in urinary cytologic samples: an outcomes and cytomorphologic study. Cancer Cytopathol. 2013;121:21–8.
14. McCroskey Z, Kliethermes S, Bahar B, Barkan GA, Pambuccian SE, Wojcik EM. Is a consistent cytologic diagnosis of low-grade urothelial carcinoma in instrumented urinary tract cytologic specimens possible? A comparison between cytomorphologic features of low-grade urothelial carcinoma and non-neoplastic changes shows extensive overlap, making a reliable diagnosis impossible. J Am Soc Cytopathol. 2015;4:90–7.
15. Xin W, Raab SS, Michael CW. Low-grade urothelial carcinoma: reappraisal of the cytologic criteria on ThinPrep. Diagn Cytopathol. 2003a;29:125–9.
16. Reid MD, Osunkoya AO, Siddiqui MT, Looney SW. Accuracy of grading of urothelial carcinoma on urine cytology: an analysis of interobserver and intraobserver agreement. Int J Clin Exp Pathol. 2012;5:882–91.
17. Ainechi S, Pambuccian SE, Wojcik EM, Barkan G. Cytomorphologic features and differential diagnosis of neoplasms with small cell features in liquid-based urinary tract specimens. J Am Soc Cytopathol. 2015;4:295–306.
18. Bardales RH, Pitman MB, Stanley MW, Korourian S, Suhrland MJ. Urine cytology of primary and secondary urinary bladder adenocarcinoma. Cancer Cytopathol. 1998;84:335–43.
19. Zardawi IM, Duncan J. Evaluation of a centrifuge method and thin-layer preparation in urine cytology. Acta Cytol. 2003;47:1038–42.
20. Nassar H, Ali-Fehmi R, Madan S. Use of ThinPrep monolayer technique and cytospin preparation in urine cytology: a comparative analysis. Diagn Cytopathol. 2003;28:115–8.

21. Piaton E, Faÿnel J, Hutin K, Ranchin MC, Cottier M. Conventional liquid-based techniques versus Cytyc Thinprep processing of urinary samples: a qualitative approach. BMC Clin Pathol. 2005;5:9.
22. Koss LG, Melamed MR. Koss' diagnostic cytology and its histopathologic bases. 5th ed. Philadelphia: Lippincott; 2006, Chapter 22, pp. 738–76.
23. Glatz K, Willi N, Glatz D, Barascud A, Grilli B, Herzog M, Dalquen P, Feichter G, Gasser TC, Sulser T, Bubendorf L. An international telecytologic quiz on urinary cytology reveals educational deficits and absence of a commonly used classification system. Am J Clin Pathol. 2006;126:294–301.
24. Hwang EC, Park SH, Jung SI, Kwon DD, Park K, Ryu SB, Park CS. Usefulness of liquid-based preparation in urine cytology. Int J Urol. 2007;14:626–9.
25. Hoda RS. Non-gynecologic cytology on liquid-based preparations: a morphologic review of facts and artifacts. Diagn Cytopathol. 2007;35:621–34.
26. Norimatsu Y, Kawanishi N, Shigematsu Y, Kawabe T, Ohsaki H, Kobayashi TK. Use of liquid-based preparations in urine cytology: an evaluation of Liqui-PREP and BD SurePath. Diagn Cytopathol. 2010;38:702–4.
27. Voss JS, Kipp BR, Krueger AK, Clayton AC, Halling KC, Karnes RJ, Henry MR, Sebo TJ. Changes in specimen preparation method may impact urine cytologic evaluation. Am J Clin Pathol. 2008;130:428–33.
28. Lu DY, Nassar A, Siddiqui MT. High-grade urothelial carcinoma: comparison of SurePath liquid-based processing with cytospin processing. Diagn Cytopathol. 2009;37:16–20.
29. Laucirica R, Bentz JS, Souers RJ, Wasserman PG, Crothers BA, Clayton AC, Henry MR, Chmara BA, Clary KM, Fraig MM, Moriarty AT. Do liquid-based preparations of urinary cytology perform differently than classically prepared cases? Observations from the CAP interlaboratory comparison program in nongynecologic cytology. Arch Pathol Lab Med. 2010;134:19–22.
30. Koss LG, Hoda RS. *Koss's* cytology of the urinary tract with histopathologic correlations. New York: Springer; 2011.
31. Huysentruyt CJ, Baldewijns MM, Rüland AM, Tonk RJ, Vervoort PS, Smits KM, van de Beek C, Speel EJ. Modified UroVysion scoring criteria increase the urothelial carcinoma detection rate in cases of equivocal urinary cytology. Histopathology. 2011;58:1048–53.
32. Raisi O, Magnani C, Bigiani N, Cianciavicchia E, D'Amico R, Muscatello U, Ghirardini C. The diagnostic reliability of urinary cytology: a retrospective study. Diagn Cytopathol. 2012;40:608–14.
33. Piaton E, Decaussin-Petrucci M, Mege-Lechevallier F, Advenier AS, Devonec M, Ruffion A. Diagnostic terminology for urinary cytology reports including the new subcategories "atypical urothelial cells of undetermined significance" (AUC-US) and "cannot exclude high grade" (AUC-H). Cytopathology. 2014;25:27–38.

Gastrointestinal Tract Cytology

4

Introduction

The gastrointestinal tract (GIT) comprises of hollow upper and lower tracts, as well as associated solid organs. Cytology of the GIT is performed to detect neoplastic and nonneoplastic lesions and often includes concurrent procurement of needle core biopsies. Sites amenable to cytological sampling include the esophagus, stomach, duodenum, pancreatobiliary tract, and liver from the upper GIT and the rectum and anus from the lower GIT.

Carcinomas are by far the most common malignancy of the GIT. Squamous cell carcinomas are usually encountered in the proximal esophagus and anus, while adenocarcinomas are more common at other GIT sites. Other primary epithelial and non-epithelial neoplastic lesions include hematopoietic, neuroendocrine, and mesenchymal (e.g., gastrointestinal stromal) tumors. Regional lymph nodes may show any infectious, benign, or malignant (either primary or metastatic) diseases. Metastases to the hollow organs of the GIT are uncommon. GIT cytology may also be performed to investigate possible inflammatory conditions such as pancreatitis or sclerosing cholangitis, metaplastic lesions such as Barrett esophagus, and benign entities such as gastric ulcer and ampullary adenoma.

The rendering of an accurate cytological diagnosis in GIT specimens is critical, and inaccurate interpretation may have serious clinical implications. Cytological diagnoses are based on cytomorphological criteria, in conjunction with the results of ancillary studies such as special stains, immunocytochemistry, flow cytometry, and molecular analyses as well as clinical and imaging data.

Cytological Reporting Guidelines

The Papanicolaou Society of Cytopathology (PSC) guidelines for pancreaticobiliary cytology stratify the risk of malignancy associated with diagnostic categories. This can help guide management. These guidelines provide diagnostic categories and criteria and also describe techniques for obtaining specimens, ancillary testing, and patient

R.S. Hoda et al., *Diagnostic Liquid-Based Cytology*,
DOI 10.1007/978-3-662-53905-7_4, © Springer-Verlag GmbH Germany 2017

follow-up and management. A six-tiered system is recommended for the standardized nomenclature for reporting pancreaticobiliary cytological diagnoses. The categories are nondiagnostic, negative, atypical, neoplastic (benign or other), suspicious, and positive. The "neoplastic (benign and other)" category includes benign cystic neoplasms (serous cystadenoma), premalignant mucinous cysts [cystic mucinous neoplasm and intraductal papillary mucinous neoplasm (IPMN)] without unequivocal features of malignancy, low-grade well-differentiated neuroendocrine tumors (NETs), and solid pseudopapillary neoplasms. A multidisciplinary approach is recommended for diagnosis. Patient management should be determined by correlating the clinical findings in concert with endoscopic, imaging, and cytological findings.

Indications, Collection, and Laboratory Processing of Cytological Samples

The examination of GIT cytology is an efficient and cost-effective method for diagnosis. It allows rapid interpretation of specimens using on-site FNA adequacy assessment, which in turn allows for specimen triage for various ancillary tests. An accurate diagnosis relies on procuring an adequate sample, optimal preparation, and expertise in interpretation.

The principal indications for the use of cytology in GIT disorders are as follows:

1. Diagnosis of neoplastic and nonneoplastic lesions
2. Routine monitoring of patients with bile and pancreatic duct strictures

Methods of Specimen Collection

The principal methods of specimen collection are:

- Endoscopic retrograde cholangiopancreatography (ERCP)
- Brushings

ERCP uses a combination of luminal endoscopy and fluoroscopic imaging. It is both an assessment and therapeutic technique for biliary and pancreatic obstruction secondary to stones, strictures, and pancreatitis. ERCP is the traditional and simplest method of obtaining a cytology specimen from a biliary stricture. It is also the most useful tool for providing relief of obstructive jaundice by biliary stent placement.

Brushings are usually performed for esophageal and gastric lesions, as well as for pancreaticobiliary duct strictures. The selection of specimen collection method and processing depends on clinical circumstances and goal of examination. Brushings have a better yield and accuracy than bile aspiration and directed focal biopsies as they can sample a bigger area of the stricture surface.

Cytopathology Laboratory Processing of GIT Specimens

ERCP specimens and brushings are usually processed as a single LBP. An accurate cytological assessment primarily requires an adequate and well-prepared sample. Please see Chap. 1 also.

Advantages of Cytological Specimens for GIT over Core Biopsy

The advantages of cytology preparations over needle core biopsies include the following:

- Sampling of larger area of concern such as in Barrett esophagus by brushings
- Sampling of "narrow" areas such as ducts and strictures in biliary and pancreatic ducts by brushings
- Ability to assess pancreatic lesions; subepithelial, submucosal, and mural nodules; and regional lymph nodes via EUS-FNA
- Collection of cyst fluid for chemical and molecular analyses
- Shorter turnaround time

Endoscopic Retrograde Cholangiopancreatography (ERCP)

ERCP is the traditional and simplest method of obtaining a cytology specimen of a biliary stricture, typically via brushing.

Brushings

Brushings are obtained through the biopsy channel of the endoscope. LBP specimens are usually processed as a single Papanicolaou (Pap)-stained slide. The collected specimen is rinsed in the preservative medium. Direct smears from brushings can also be made with a quick rolling motion of the brush on glass slides. The slides can be fixed in 95 % ethanol for Pap staining or air-dried for Romanowsky staining (e.g., Diff-Quik). Residual material can be rinsed in collection medium and used to process LBP or cell block preparation if tissue fragments are noted.

Esophageal brushings are usually performed to exclude Candida infection and occasionally for esophageal malignancy.

Gastric brushings are usually performed to distinguish benign from malignant gastric ulcers or to sample a gastric mass.

Rectal brushings may be performed on a rectal mass to rule out malignancy.

Bile duct brushings are usually performed to determine the cause of a stricture.

Pancreatic duct brushings are usually performed to determine the cause of a stricture.

Anal Pap tests are usually performed in immunocompromised patients to exclude anal intraepithelial neoplasia (AIN) or malignancy.

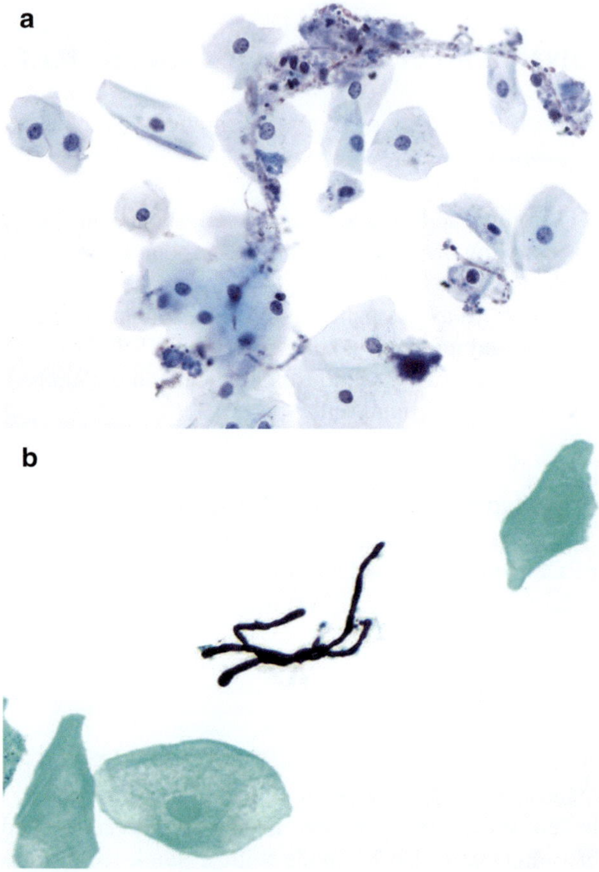

Fig. 4.1 Esophageal brushings with fungal organisms. (**a**) The fungal organism depicted here is composed of yeast and pseudohyphae which impale reactive squamous cells, giving a "shish kebab" appearance. These findings are most compatible with *Candida spp*. Adjacent squamous cells show mild reactive changes, i.e., a slight increase in nuclear size, but the chromatin is bland and the nuclear borders are regular. The background is clean (TP). (**b**) When a structure is suspected to be of fungal origin, a silver stain such as Grocott's methenamine stain (GMS) can be performed on additional LBP. Most fungal yeast and hyphae are positive by GMS. Occasionally, GMS may stain normal background elements (GMS on TP)

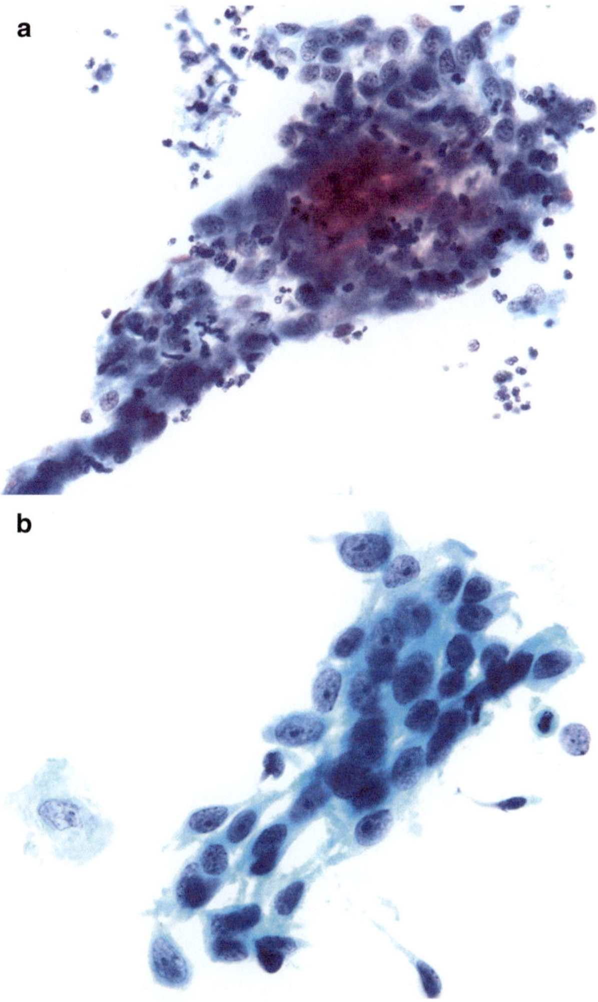

Fig. 4.2 Esophageal brushings with reactive changes. (**a**) The squamous cells are cohesive and form two-dimensional flat sheet with smooth borders and relative cell polarity. The cells are enlarged and uniform, with expanded vesicular to pinkish-staining nuclei that have thin, smooth membranes, and single or multiple micronucleoli. Mitoses were present elsewhere. Cytoplasm shows occasional vacuoles and two-toned eosinophilic and basophilic staining. The background shows inflammatory cells. Nuclear-to-cytoplasmic ratio is low. Reactive changes can be secondary to infections, trauma, inflammation, radiation, and chemotherapy. In treatment-related changes, secondary to radiation and chemotherapy, cell enlargement (cytomegaly) with two-tone cytoplasmic staining and cytoplasmic vacuolization may be seen, as demonstrated by the sheets of cells. Cell enlargement is usually secondary to increased cytoplasm; therefore nuclear-to-cytoplasmic ratio is not elevated. Wang et al. [1] noted improved overall quality of the TP endoscopic brushing preparations, with an even distribution of cells, less obscuring blood and inflammatory cells, and air-drying artifact, than compared to direct smears. (**b**) In contrast to reactive changes, squamous cell carcinoma shows two-dimensional cell sheets with irregular borders. The nuclei therein are slightly irregular, enlarged with coarser clumped chromatin, macronucleoli, and parachromatin clearing. Cytoplasm appears dense and rigid with bipolar extensions. Nuclear-to-cytoplasmic ratio is high. Note the highly atypical spindled or elongated tumor cells. (**a** and **b**, TP)

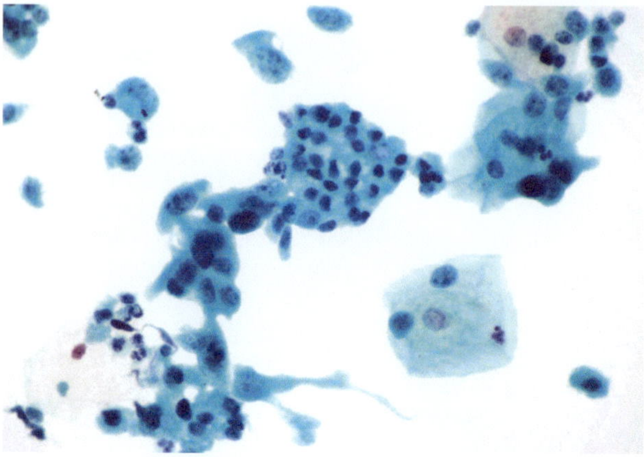

Fig. 4.3 Esophageal brushings with Barrett esophagus. Compare the glandular cells in this figure to the squamous cells found in the previous two figures. These brushings are from Barrett esophagus. Note squamous cells in apparent continuity with a honeycombed sheet of glandular cells, a characteristic cytomorphological feature of Barrett esophagus. Also note a goblet cell containing a pale-staining mucin globule in the honeycombed sheet of glandular cell. Intestinal metaplasia, as evident by the presence of goblet cells, is characteristic of Barrett esophagus. Some cells next to the honeycomb sheets show dysplastic features. This patient subsequently developed adenocarcinoma in Barrett esophagus (TP). Walavalkar et al. [2] found that TP LBP of distal esophageal brushings had good sensitivity (82 %) and specificity (88 %) for identifying intestinal metaplasia and high-grade dysplasia (HGD) compared with biopsy

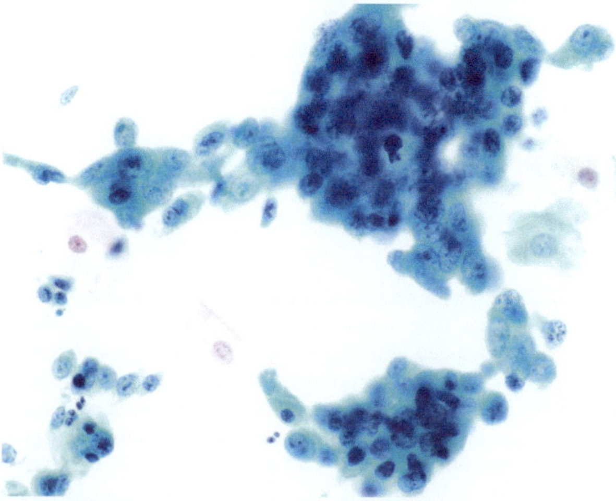

Fig. 4.4 Esophageal brushings with esophageal adenocarcinoma. Same case as Fig. 4.3 demonstrates a large number of malignant glandular cells. The cells are columnar, with delicate cytoplasm, suggesting their glandular origin. The malignant nature of these cells is indicated by enlarged nuclei with coarse chromatin and irregular nuclear borders. In some cells, there are prominent nucleoli. The nuclear-to-cytoplasmic ratio is high, and there is a great variation in nuclear size. The cells form three-dimensional structures. Single malignant cells are present. Scattered benign squamous cells are also seen. High-grade dysplasia in Barrett's epithelium cannot be reliably distinguished from adenocarcinoma. However, the College of American Pathologists (CAP) Interlaboratory Comparison Program found that TP preparations of esophageal brushings performed superior to conventional smears in detecting malignancies, especially adenocarcinoma [3] (TP)

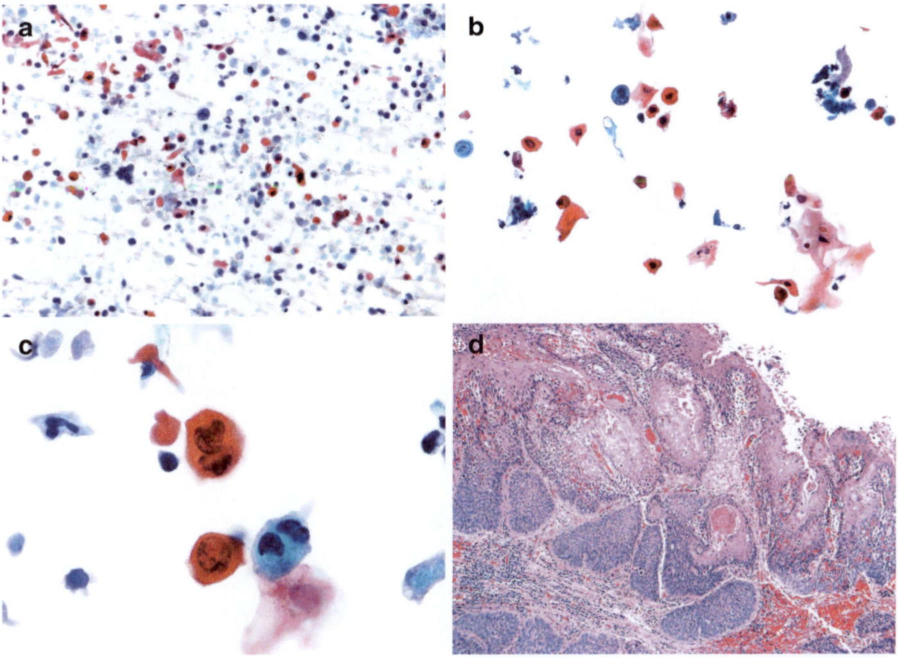

Fig. 4.5 Esophageal brushings with esophageal keratinizing squamous carcinoma. (**a**) A direct smear shows relatively large (compared to lymphoid cells) and singly dispersed tumor cells with keratinized cytoplasm. The keratinized cytoplasm is densely orangeophilic to opaque and is pleomorphic with rigid irregular extensions. The nuclei are hyperchromatic (India ink-like) with dense convoluted borders. Nuclear-to-cytoplasmic ratio is low. Background contains tumor diathesis and fragments of keratinaceous debris. Esophageal brushing cytology cannot reliably distinguish squamous dysplasia from invasive carcinoma, though abundant necrosis would be unusual for a noninvasive carcinoma (DS). (**b**) Same case on TP shows similar findings, except cellularity is lower than compared to direct smear. Note cytoplasmic and nuclear similarity to direct smear. Tumor diathesis appears clumped. Differential diagnosis of squamous carcinoma and squamous dysplasia includes reactive changes (TP) (see Fig. 4.2a). (**c**) A higher-power view of malignant keratinized squamous cells in which the nuclei appear both pale and hyperchromatic. The dense keratin often prevents the nuclei from staining darkly in some cells. Note rigid, hard cytoplasmic borders. Diathesis and keratinaceous debris is evident (TP). (**d**) Histology of the corresponding infiltrative squamous carcinoma (H&E)

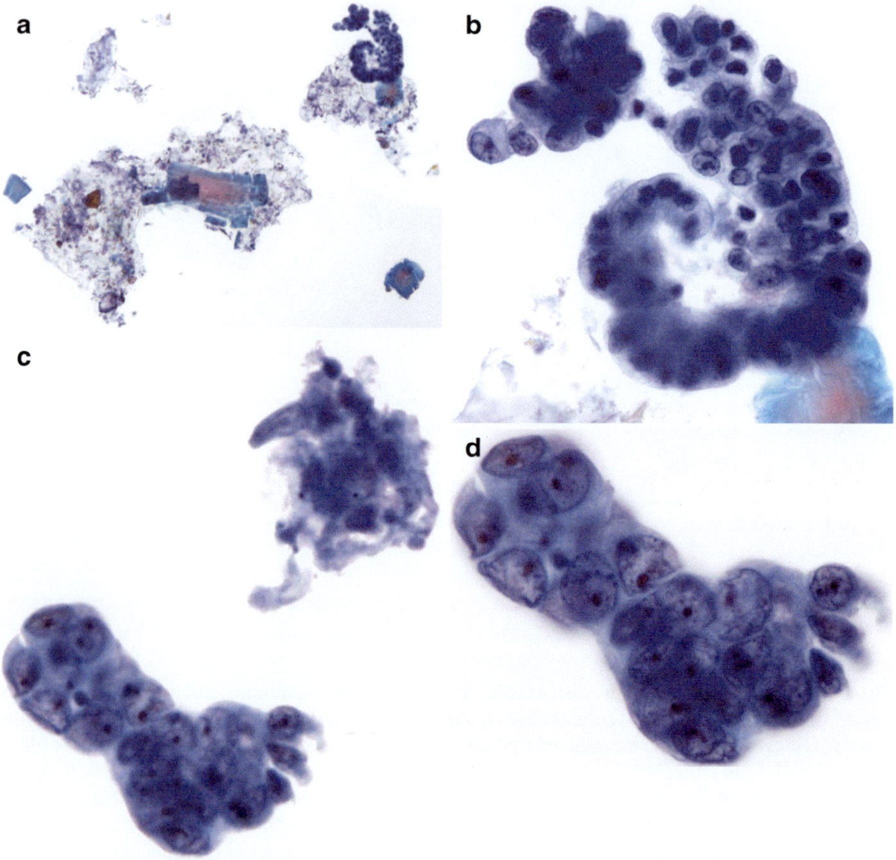

Fig. 4.6 Pancreatic duct stent and brushing. (**a**) Pancreatic duct stent in a patient with a stricture shows clumps of acellular debris, mucinous material, fungal spores, and clustered ductal cells. This is the typical background for a stent specimen (TP). (**b**) The cellular cluster shows stent-associated reactive atypia. The cluster has rounded borders, with retained polarity. The cells have relatively low nuclear-to-cytoplasmic ratio; round to oval, vesicular nuclei; and prominent single nucleoli. Cytoplasm is finely vacuolated. No single cells or loosely cohesive clusters are noted (TP). (**c**) Pancreatic duct brushings in a case of ductal adenocarcinoma of pancreas shows a malignant gland with focal irregularity and clumped diathesis with some embedded tumor cells (TP). (**d**) The malignant gland comprises of cells with high nuclear-to-cytoplasmic ratio; loss of cell polarity, with enlarged irregular nuclei with coarsely clumped chromatin; and single cherry red macronucleoli. Compare with stent-associated reactive atypia seen in **(b)** (TP)

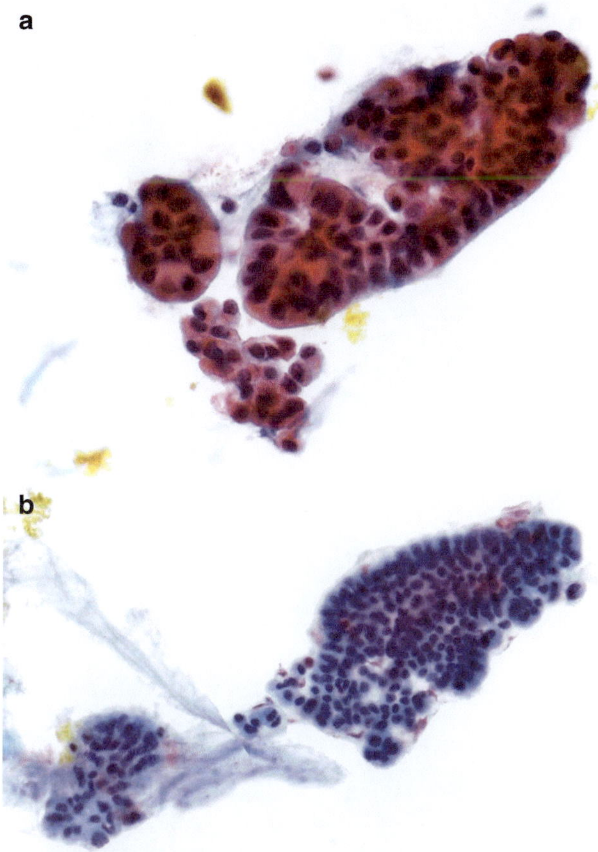

Fig. 4.7 Bile duct brushing (BDB) with benign changes. (**a**) On conventional brush cytology, normal bile duct epithelium appears as cohesive monolayer sheets of small- to medium-sized epithelial cells with centrally located round to oval nuclei and low nuclear-to-cytoplasmic ratio. The nuclei are uniform with finely granular evenly distributed chromatin and inconspicuous nucleoli (Pap-stained DS). BDB is an important diagnostic modality to exclude pancreatobiliary malignancy (TP). LBP has been shown to have similar sensitivity and specificity as direct smear while providing better preservation and cytological detail [4]. (**b**) In TP, the cytomorphology is similar, except that nuclei may appear smaller. Note bile pigment loosely associated with the tissue fragment (TP)

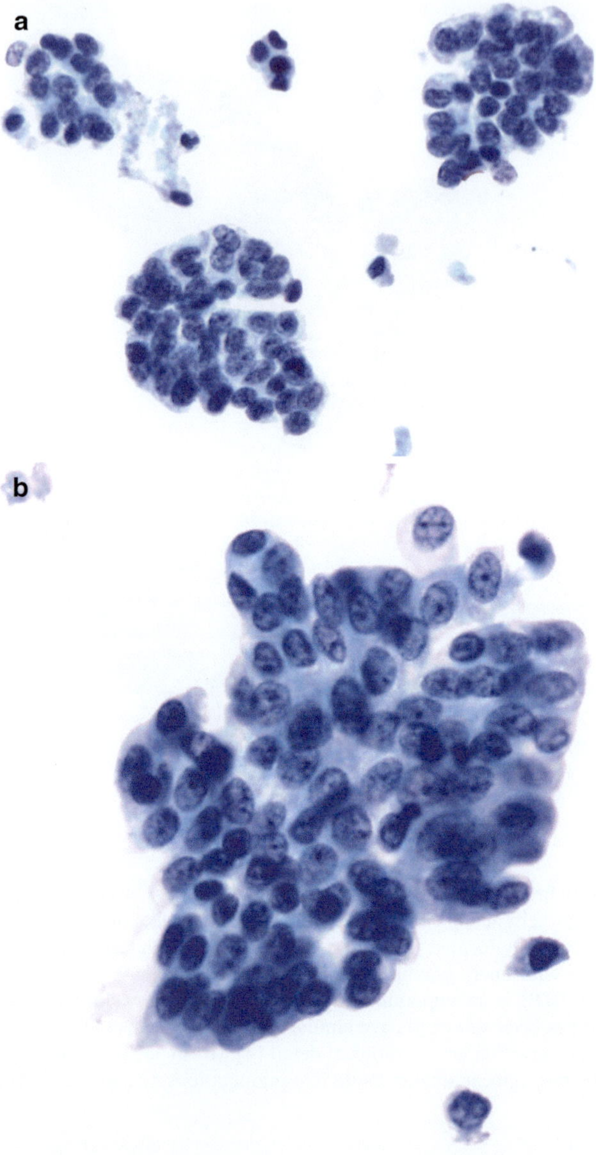

Fig. 4.8 Bile duct brushing with reactive changes. (**a**, **b**) Reactive changes in BDB may be seen secondary to cholelithiasis, cholangitis, or indwelling stent. Reactive changes include slightly increased cellularity with sheets and clusters of ductal epithelial cells. The clusters have a smooth contour or edges with a rim of cytoplasm. The reactive cells have slight nuclear overlap and slightly enlarged nuclei with smooth nuclear membranes, finely granular chromatin, and single or multiple micronucleoli. Note retained cell polarity (TP). When a stent is present, nuclear size variation and nucleoli may be seen, but nuclei maintain their boundaries; the cytoplasm may be elongated and "streaming," and intracytoplasmic neutrophils may be present [5]

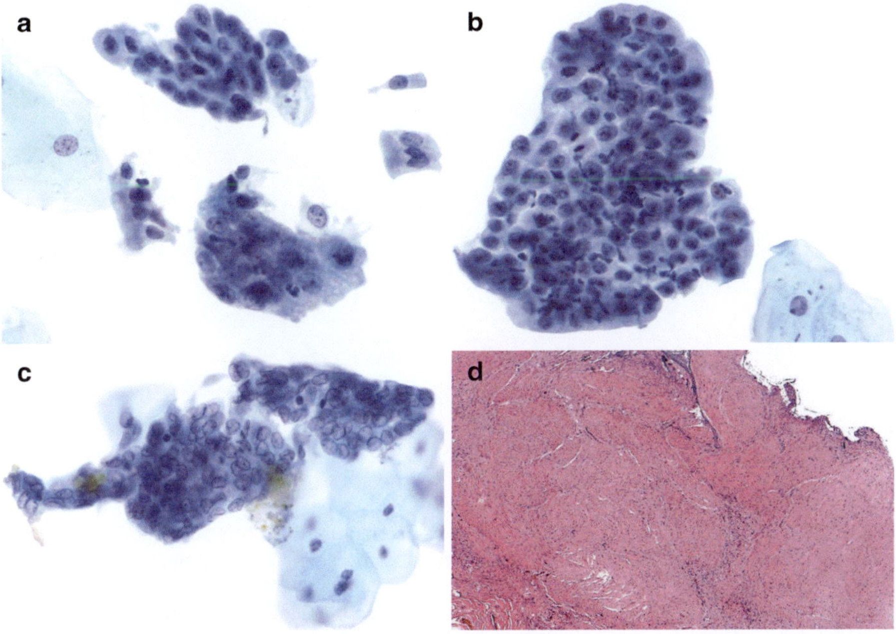

Fig. 4.9 Bile duct brushing with reactive changes secondary to stones. (**a**) In this BDB from a patient with cholelithiasis, ductal cells are seen singly, in strips with streaming effect, and as larger fragments. All have increased nuclear size, but nuclear-to-cytoplasmic ratio is not greatly increased. The cells maintain nuclear polarity and have a columnar configuration. Some cells have prominent nucleoli, but nuclear borders remain regular (TP). (**b**) This fragment contains reactive ductal cells with prominent nucleoli. Note embedded acute inflammatory cells. Reactive changes in BDB may result in a slightly distorted honeycomb arrangement (TP). (**c**) This group shows degenerative changes. Although the cells show slight nuclear crowding and loss of polarity and a moderate increase in nuclear-to-cytoplasmic ratio, these are reactive/degenerative changes. The chromatin is fine and pale, and nucleoli are absent. BDB cytology is unreliable for detecting LGD, while HGD is usually diagnosed as suspicious for adenocarcinoma (see description below) (TP). (**d**) Resection specimen shows denuded epithelium with underlying thickened connective tissue. Gallstones were present in this case (H&E)

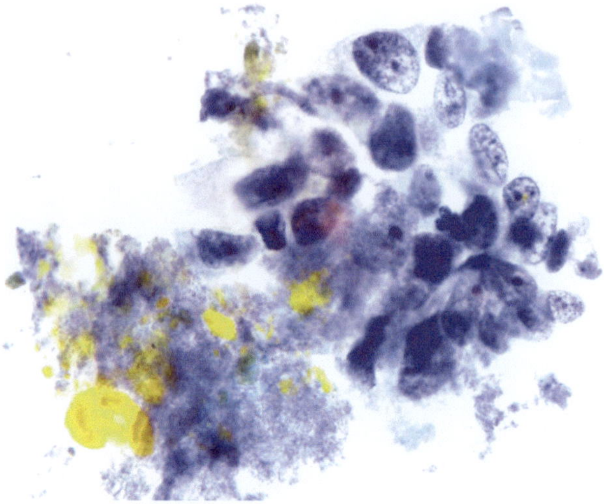

Fig. 4.10 Bile duct brushing, suspicious for adenocarcinoma. BDB of a stricture. The enlarged cells have coarse chromatin and demonstrate anisonucleosis. The enlarged nuclei are elongated with variable shape and show prominent cherry red nucleoli with parachromatin clearing and occasional grooves. It is difficult to assess the cytoplasm. The cells are associated with bile pigment and granular debris. This amount of atypia is suspicious for adenocarcinoma – even with the knowledge of stent placement (TP). A bile duct stricture can result in reactive atypia and mimic adenocarcinoma. Oftentimes these patients require stent placement. Stents contribute to reactive atypia. Therefore, it is important to know patient history and be cautious with diagnostic interpretation. In one study of 184 BDB diagnosed as suspicious, almost 80 % had malignancy on follow-up [6]

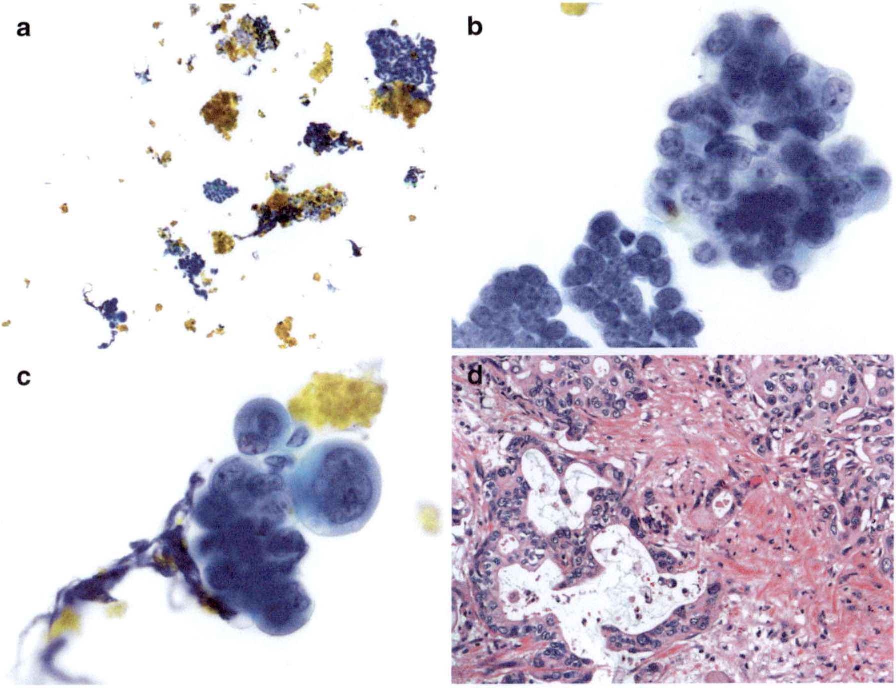

Fig. 4.11 Bile duct brushing, pancreatic adenocarcinoma. (**a**) BDB of a stricture: cellular specimen with small to large HCGs with intermixed bile pigment. The groups have irregular outlines. It is difficult to assess atypia on low magnification (TP). (**b**) Malignant cells appear in a cluster with nuclear overlapping and crowding. The cells have high nuclear-to-cytoplasmic ratio, with irregular nuclear membranes, coarse chromatin, creases and grooves, prominent nucleoli with parachromatin clearing (TP). Henke et al. [7] reviewed 90 BDB from 80 patients on DS and identified three primary cytological features that were most frequently useful in distinguishing biliary tract malignancies from benign strictures. The primary criteria were increased nuclear-to-cytoplasmic ratio, nuclear molding, and chromatin clumping. Salient secondary criteria included disordered cellular sheets with altered cell polarity and nuclear features of enlargement, anisonucleosis, irregular membranes, grooves, macronucleoli, and chromatin clearing. The authors concluded that the presence of two of three primary criteria resulted in sensitivity of 83 %. Minamiguchi et al. [8] reported the most specific features to be mitoses, clumped chromatin, nuclear membrane irregularity, cohesiveness, and enlargement of nuclei in BDB specimens. These studies have shown that sensitivity rates for BDB specimens for TP and DS are comparable. (**c**) A small HCG of malignant cells. Note loss of polarity and distinct cytoplasmic borders. The nuclear size variation is striking (TP). (**d**) The histology of pancreatic adenocarcinoma on resection demonstrates crowded and disorganized glands with an infiltrative pattern; the nuclear features are similar to those seen in **(a–c)** (H&E)

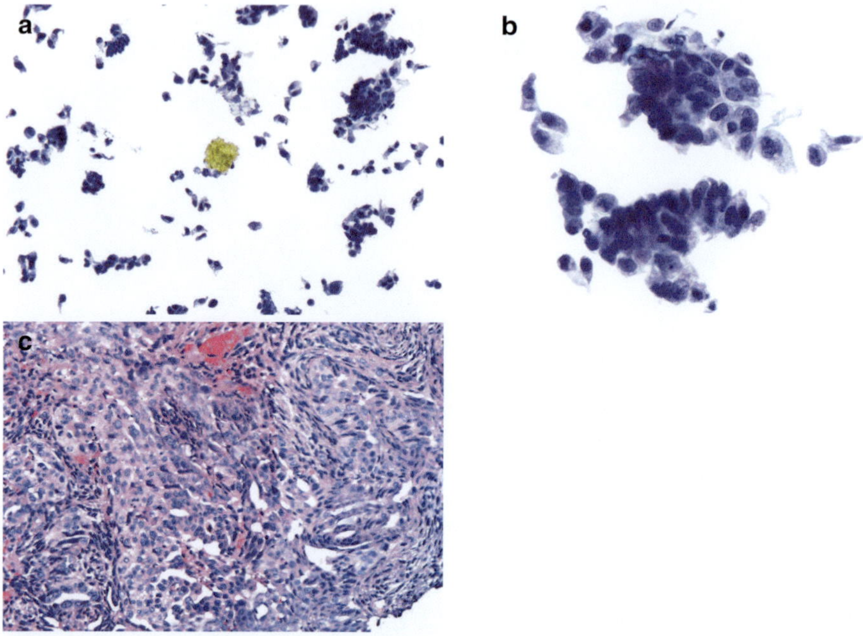

Fig. 4.12 Bile duct brushing with cholangiocarcinoma. (**a**) Here a cholangiocarcinoma demonstrates a hypercellular specimen with small to large hyperchromatic crowded groups, sheets, strips, and single malignant cells with intermixed bile pigment and clumped necrosis. The cells are predominantly columnar and appear hyperchromatic with irregular nuclear shapes and sizes (TP). (**b**) The malignant features are similar to those described for pancreatic adenocarcinoma. Here, the cells maintain a columnar appearance, but nuclei are dark and overlapping. In some foci, nuclear polarity is lost (TP). (**c**) Cholangiocarcinoma on concurrent biopsy (H&E)

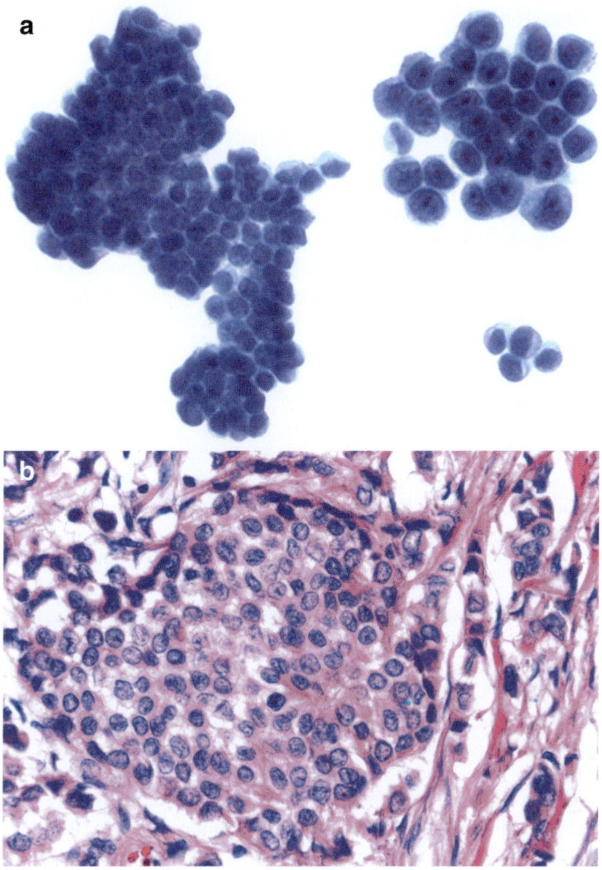

Fig. 4.13 Bile duct brushing with metastatic carcinoma. Lobular carcinoma of the breast metastatic to the bile duct: (**a**) the cluster of cells (on the right) represent lobular carcinoma metastatic to the bile duct. For metastases to be present on a BDB, the tumor must have eroded into bile duct lumen. In this case, the carcinoma cells are monotonous with round eccentrically placed nuclei and a minute rim of cytoplasm. In this case, the cells have prominent nucleoli. Benign cells are also present (on the left). Tumor cells are larger and slightly disorganized than benign cells. It is important to know the patient's history to exclude a primary neoplasm. If cell block material is available, immunostains for breast-specific markers can be confirmatory (TP). (**b**) the corresponding histology of the resected specimen shows tumor cells to be nested, with a monotonous appearance. Individual infiltrating cells can also be seen (H&E)

Suggested Reading

1. Wang HH, Trawinski SS, Garcia LW, Abu-Jawdeh GM, Upton M, Werneke S. ThinPrep processing of endoscopic brushing specimens. Am J Clin Pathol. 1996;105:163–7.
2. Walavalkar V, Patwardhan RV, Owens CL, Lithgow M, Wang X, Akalin A, Nompleggi DJ, Zivny J, Wassef W, Marshall C, Levey J, Walter O, Fischer AH. Utility of liquid-based cytologic examination of distal esophageal brushings in the management of Barrett esophagus: a prospective study of 45 cases. J Am Soc Cytopathol. 2015;4:113–21.
3. Clayton AC, Bentz JS, Wasserman PG, et al. Comparison of ThinPrep preparations to other preparation types in gastrointestinal cytology: observations from CAP interlaboratory comparison program in non-gynecologic cytology. CAP cytopathology resource committee. Arch Pathol Lab Med. 2010;134:1116–20.
4. Siddiqui MT, Gokaslan ST, Saboorian MH, Carrick K, Ashfaq R. Comparison of ThinPrep and conventional smears in detecting carcinoma in bile duct brushings. Cancer. 2003;99:205–10.
5. Chadwick BE. Beyond cytomorphology: expanding the diagnostic potential for biliary cytology. Diagn Cytopathol. 2012;40:536–41.
6. Volmar KE, Vollmer RT, Routbort MJ, Creager AJ. Pancreatic and bile duct brushing cytology in 1000 cases. Cancer Cytopathol. 2006;108:231–8.
7. Henke AC, Jensen CS, Cohen MB. Cytologic diagnosis of adenocarcinoma in biliary and pancreatic duct brushings. Adv Anat Pathol. 2002;9:301–8.
8. Minamiguchi S, McEvoy R, Fraig M, Lewin DN, Wallace MB, Hoda RS. Bile duct brushings on ThinPrep®. Diagn Cytopathol. 2004;30:292–3.
9. Young JA, Elias E. Gastro-oesophageal candidiasis: diagnosis by brush cytology. J Clin Pathol. 1985;38:293–6.
10. Nguyen M, Mikita G, Hoda RS. "Intercellular bridges" in a case of well differentiated squamous carcinoma. Diagn Cytopathol. 2016;44:121–3.
11. Koybasioglu F, Onal B, Simsek GG, Yilmazer D, Han U. Comparison of ThinPrep and conventional smears in head and neck fine needle aspiration cytology. Turkish J Pathol. 2008;24:159–65.
12. Conrad R, Castelino-Prabhu S, Cobb C, Raza A. Role of cytopathology in the diagnosis and management of gastrointestinal tract cancers. J Gastrointest Oncol. 2012;3:285–98.
13. West RB, Corless CL, Chen X, Rubin BP, Subramanian S, Montgomery K, Zhu S, Ball CA, Nielsen TO, Patel R, Goldblum JR, Brown PO, Heinrich MC, van de Rijn M. The novel marker, DOG1 is expressed ubiquitously in gastrointestinal stromal tumors irrespective of KIT or PDGFRA mutation status. Am J Pathol. 2004;165:107–13.
14. Hoda R. Non-gynecologic cytology on liquid-based preparations: a morphologic review of facts and artifacts. Diagn Cytopathol. 2007;35:621–34.

Body Cavity Fluids

5

Introduction

The body cavities, including pleural, pericardial, and peritoneal cavities, lie within a double-layered serous membrane lined by flat mesothelial cells. The inner layer invests the organs and is called the visceral layer, and the outer is called the parietal layer. A potential space separates the two layers. Under normal conditions the cavities contain only minimal amount fluid which lubricates the two adjacent layers as they move. Larger amount of fluid, an effusion, accumulates during disease states.

Two types of effusions are recognized, transudate and exudate.

- *Transudate* results from imbalance of hydrostatic and oncotic pressures. Hydrostatic pressure is increased and oncotic pressure is reduced in congestive heart failure, cirrhosis, peritoneal dialysis, and nephrotic syndrome. Transudate may be straw-colored, clear or opalescent, and watery, with a low protein content of <3 g/dL, low lactate dehydrogenase (LDH), and specific gravity of less than or equal to 1.015 with low cellularity.
- *Exudate* results from increased capillary permeability due to injury to mesothelium as in malignancy, inflammatory conditions, connective tissue diseases, pulmonary infarction, drug sensitivity, or trauma. Exudates have relatively high total protein content of >3 g/dL, high LDH, and a specific gravity of more than 1.015 with high cellularity.

The distinction between transudate and exudate is made by measurement of protein concentration and specific gravity. This distinction is important because cytological examination of a transudate is generally not needed, whereas an exudate may result from malignant tumors or infectious processes and requires cytological assessment.

R.S. Hoda et al., *Diagnostic Liquid-Based Cytology*,
DOI 10.1007/978-3-662-53905-7_5,

Body Cavity Fluid Preparations

TP and SP have been utilized for non-gynecologic (non-gyn) specimens since 1991 and 1999, respectively. Since then, the use of LBP has become widespread. Several laboratories have now substituted traditional preparations (i.e., smears, filters, cytocentrifuges, and cell block) with LBP or now use LBP in addition to the classical methods. LBP perform as well, and sometimes better than, traditional preparations.

Types of Body Cavity Fluid

The body cavity fluid specimens pose a daily challenge in current cytopathology practice, especially with regard to distinguishing malignancies from reactive mesothelial cells. Specimen types include pleural, peritoneal (ascites), and pericardial effusions, cerebrospinal fluid, and pelvic washings (PW). Neoplastic entities can be:

1. *Pleural and peritoneal effusions*
   - *Primary*
     - Mesothelioma
     - Papillary serous carcinoma (peritoneal effusion)
   - *Secondary* (*metastatic*)
     - Epithelial
       Adenocarcinoma of the lung, breast, GIT, and gynecological origin
       Squamous cell carcinoma
       Small cell carcinoma
     - Non-epithelial
       Hematopoietic and lymphoid malignancies
       Melanoma
       Sarcoma
2. *Pelvic washings*
   - Same as the above

Immunocytochemistry

Immunocytochemistry is useful in distinguishing reactive mesothelial cells from malignant cells, evaluation of unknown primary sites of origin, and confirming a known malignancy involving body cavity fluids. For immunostaining, cell block sections are recommended, but immunostains can also be performed on additional LBP made from residual specimens.

Cytology of Body Cavity Fluids on LBP

The cytological criteria of malignancy include high specimen cellularity with two distinct cell populations. In a CAP interlaboratory comparison program TP performed slightly better than classical preparations in diagnosing adenocarcinoma in body cavity fluid cytology. In this regard, some caveats follow:

- With malignant effusions, typically there is a history of malignancy.
- An effusion as primary presentation of malignancy is rare.
- Bloody effusions are more likely to be associated with malignancy (blood does not obscure cells in LBP).
- Malignant effusions show high cellularity and cellular discohesion.
- Pleural effusions, processed as TP, do not appear to provide additional diagnostic value when compared to cytospin DQ-stained preparations for distinguishing mesothelioma from adenocarcinoma, since the key distinguishing cytological features of mesothelioma and adenocarcinoma can be observed in both preparations [1].
- Malignant cells in body cavity fluids differ from those in exfoliative, brushing, and FNA specimens.
- Cells "round up" in effusions, and this feature is more prominent in LBP.

Diagnostic Categories for Body Cavity Fluid Cytology

Usually four diagnostic categories are used including negative, atypical, suspicious, and positive for malignancy.

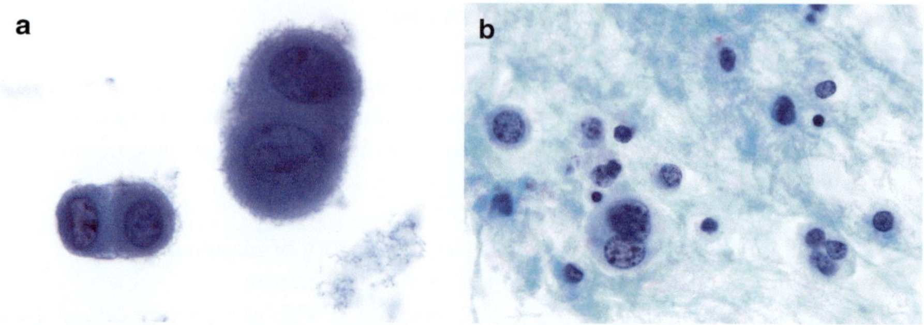

Fig. 5.1 Benign pleural effusion. (**a**) Mesothelial cells are the most common type of cells in effusions. Mesothelial cells are larger than other cells typically seen in effusions (i.e., lymphocytes and histiocytes) and contain abundant amphophilic cytoplasm. There is peripheral paler staining of cytoplasm (cytoplasmic "skirt"). Mesothelial cells may appear singly or in clusters. When two mesothelial cells lie together, they often form a "window" between the cells due to long microvilli. Epithelioid features in mesothelial cells can be mistaken for metastatic adenocarcinoma, especially when reactive changes are present (TP). (**b**) The larger cells with coarser chromatin, one binucleated, are mesothelial cells; other cells present are lymphocytes and histiocytes. While LBP typically shows a clean background, some granular material may be evident, representing degenerated blood cells and fibrin (TP)

Fig. 5.2 Benign peritoneal washing. Compared to effusions, washing specimens have a tendency to disrupt the mesothelial lining. Here, a monolayered sheet of benign mesothelial cells can be seen with an organized honeycomb appearance. The chromatin is bland. Also present are single mesothelial cells, inflammatory cells, and histiocytes. Mesothelioma should be excluded when sheets of mesothelial cells are seen in effusions (TP)

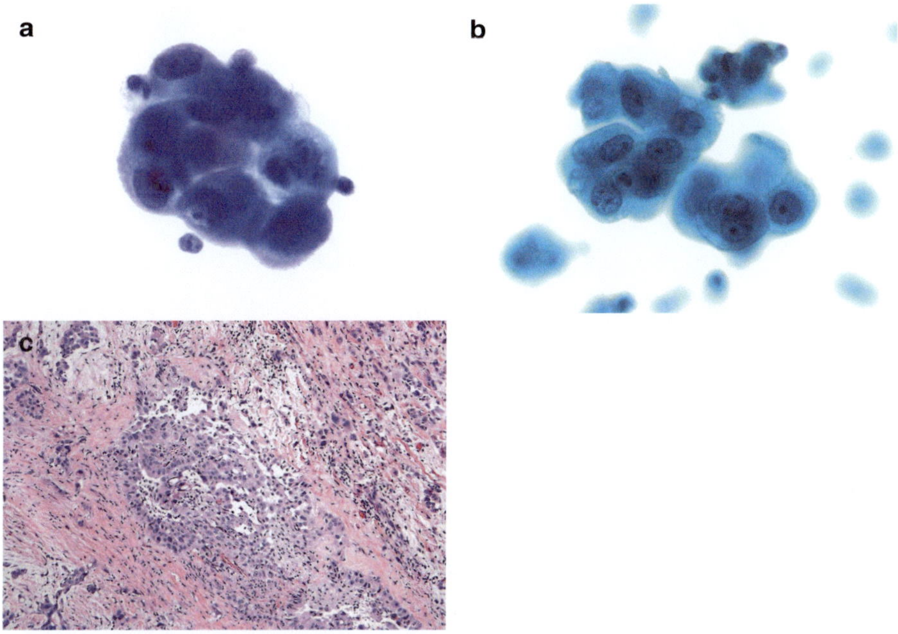

Fig. 5.3 Mesothelioma. (**a**) Compared to sheets of mesothelial cells seen in previous figures, mesothelioma cell nuclei are enlarged and overlap. Note parachromatin clearing around macronucleoli and nuclear hyperchromasia. The "hobnail" appearance is more suggestive of mesothelial origin than adenocarcinoma. Clinical history of asbestos exposure and smoking and radiological evidence of pleural or peritoneal plaques or nodularity are helpful (TP). (**b**) Mesothelioma in SP specimen. Note similarity in cytological appearance with TP in (**a**) with "hobnail" appearance, prominent nucleoli, parachromatin clearing, and "windows" between cells (SP). Mesothelioma can have different morphologies [2]. (**c**) Histologically, the mesothelial cells are epithelioid and infiltrate pleura (H&E). The images depicted are from a pleural effusion from an 80-year-old shipyard worker. Immunostains can help exclude metastatic adenocarcinoma, a malignancy which is more common in effusions than mesothelioma. Mesothelioma (and reactive mesothelial cells) is positive for calretinin and WT1, while adenocarcinoma is positive for epithelial markers such as BerEP4 and MOC31

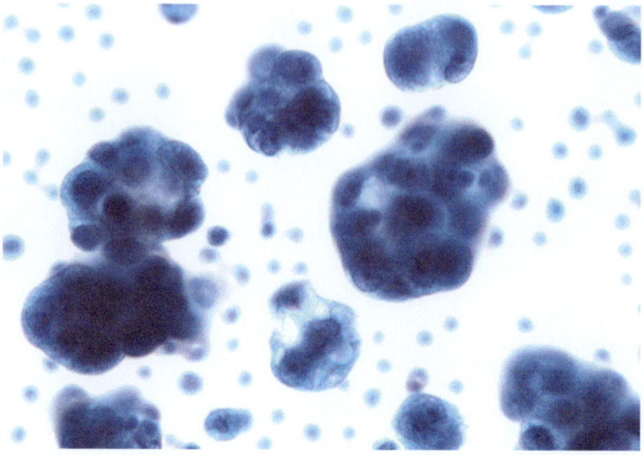

Fig. 5.4 Metastatic ovarian adenocarcinoma. The specimen is hypercellular, with clusters of malignant cells that have prominent nucleoli and high nuclear-to-cytoplasmic ratio. The carcinoma cells are large and their three-dimensional quality is maintained on SP. Some cells show mucinous vacuoles which indent nuclei. Note smooth edges ("community" border) of cell clusters. The background cells are out of focus being in a different plane of focus (SP). Reactive mesothelial cells can also have vacuolization that may be mistaken for mucin [3]; however, benign vacuoles in mesothelial cells do not indent nuclei

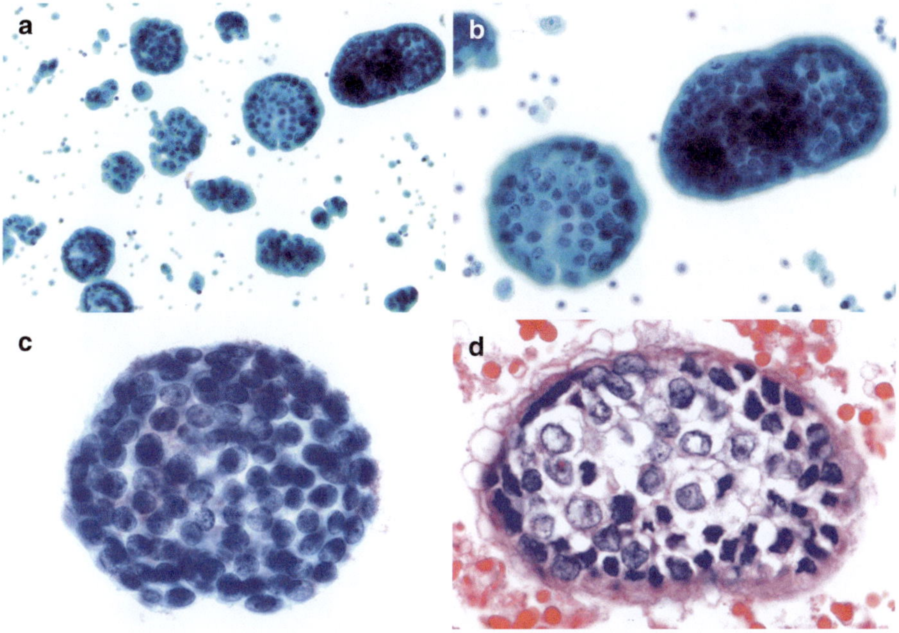

Fig. 5.5 Metastatic ductal carcinoma of the breast. (**a**, **b**) Metastatic ductal carcinoma of the breast showing high cellularity. Note three-dimensional clusters of cells with smooth borders, known as proliferation spheres or "morulas," a characteristic feature of metastatic breast carcinoma. Compare this to the scalloped border of mesothelioma in Fig. 5.3a, b. Single malignant cells are present. Nuclei are moderately enlarged, somewhat hyperchromatic with nucleoli and parachromatin clearing, and cytoplasm is denser (**a**, SP; **b**, SP). (**c**) The morula as seen on TP has similar cytomorphology to SP. However, cytoplasm is less dense than SP (TP). Most effusions are evaluated for metastatic malignancies and usually show two-cell population of benign mesothelial cells and malignant cells. **(d)** The corresponding correlate of morula on cell block (H&E)

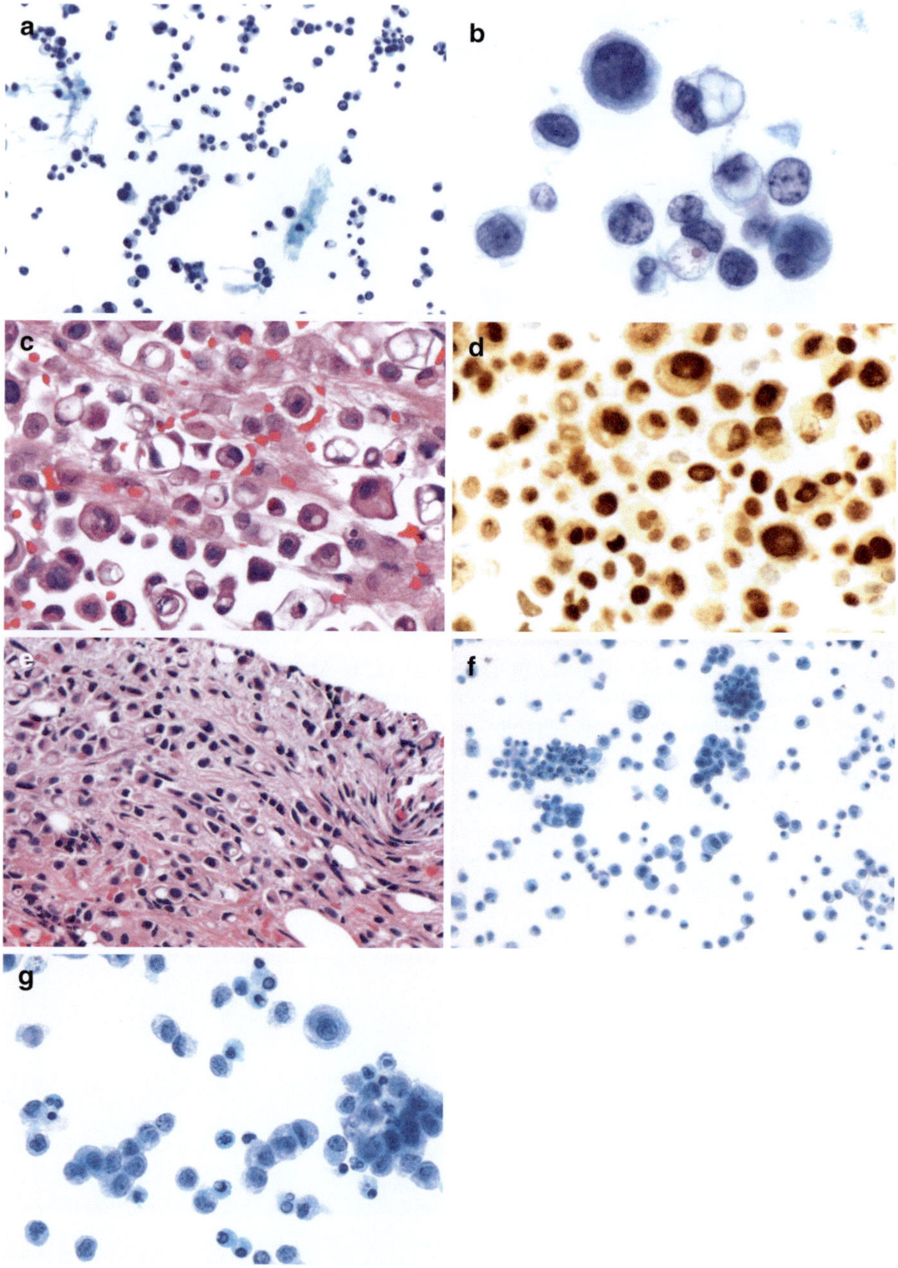
a
b
c
d
e
f
g

Fig. 5.6 Metastatic lobular carcinoma of the breast. (**a**) A highly cellular specimen, but in contrast to previous figure of ductal breast carcinoma. Lobular carcinoma cells are more discohesive and are more difficult to distinguish from background mesothelial cells. Some cells appear to form distinct linear arrangements ("Indian-file" pattern). Note small cell size and eccentrically located nuclei (TP). (**b**) Closer inspection reveals the neoplastic cells vary in size and shape (pleomorphic). Nuclei are eccentric with irregular, thick membranes, vesicular chromatin, and small nucleoli. The cytoplasm is vacuolated. Note characteristic signet ring cell with cytoplasmic vacuole and pink-staining mucinous condensation. Such cells are pathognomonic for lobular carcinoma of the breast (TP). (**c**) Cytomorphology seen on cell block is comparable to TP. A few cells show mucinous condensation, which would be highlighted by a mucicarmine stain (H&E). (**d**) An immunohistochemical study for GATA-3 on the cell block section shows strong nuclear positivity, confirming breast origin (GATA-3 IHC). One study has shown that immunostains perform equally well on TP as on cell block sections. The latter preparations are superior for nuclear markers such as p53 [4]. (**e**) Core biopsy of primary lobular carcinoma shows similar cytological features as TP and cell block, with infiltrating single cells (H&E). (**f**, **g**) Metastatic lobular carcinoma of the breast in a pleural effusion processed as SP. Cytomorphology is similar to that described for TP in figures (**a** and **b**) (SP)

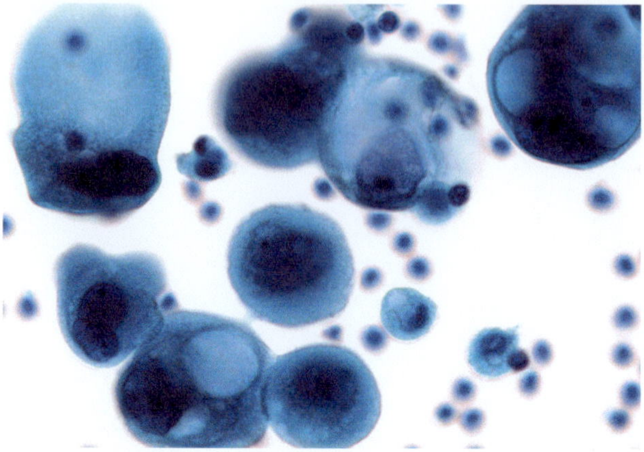

Fig. 5.7 Metastatic adenocarcinoma of the lung. This is one of the most common malignancies to involve pleural cavity. Malignant cells are greatly enlarged and show enlarged, irregular, hyperchromatic nuclei with prominent nucleoli and large cytoplasmic vacuoles, some multiple. Mucinous condensation is not seen. In a patient with a history of lung adenocarcinoma and absence of any other malignancy, the cytomorphology alone is enough for a diagnosis without confirmatory IHC. Molecular assessment, if requested by clinicians, can be performed on cell block preparations (SP)

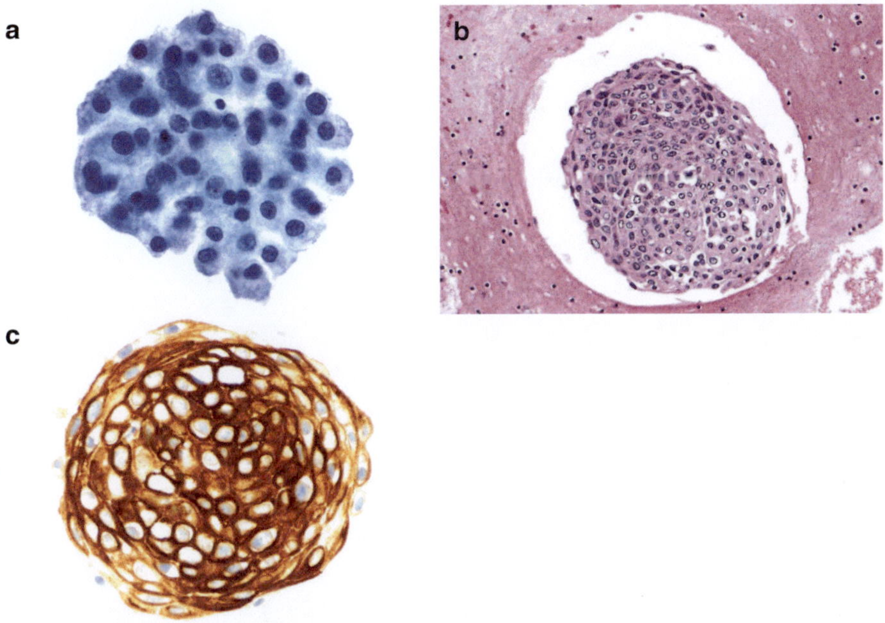

Fig. 5.8 Metastatic squamous cell carcinoma of the lung. (**a**) This was a challenging case as malignant cytology on TP was not clearly evident. The cell block preparation was instrumental in rendering an accurate diagnosis. Squamous cell carcinoma uncommonly involves pleural effusions; the cells here have centrally placed nuclei and appear to have spaces between the cytoplasmic borders. This feature may mimic "windows" seen between cells of mesothelial origin and suggest mesothelioma. The cytoplasm does appear somewhat dense, and the cell border appears smooth [5] (TP). (**b**) The cell block material shows a rounded cluster of tumor cells within a distinct lacunar space, often formed artifactually around metastatic carcinomas on cell block preparations. Even at low-power microscopy, cytology appears similar to TP (H&E); (**c**) immunostain for cytokeratin 5/6, performed on cell block sections, shows strong cytoplasmic positivity (CK5/6 IHC). Other squamous cell carcinoma markers such as p40 and p63 (nuclear stains) were also immunoreactive in this case

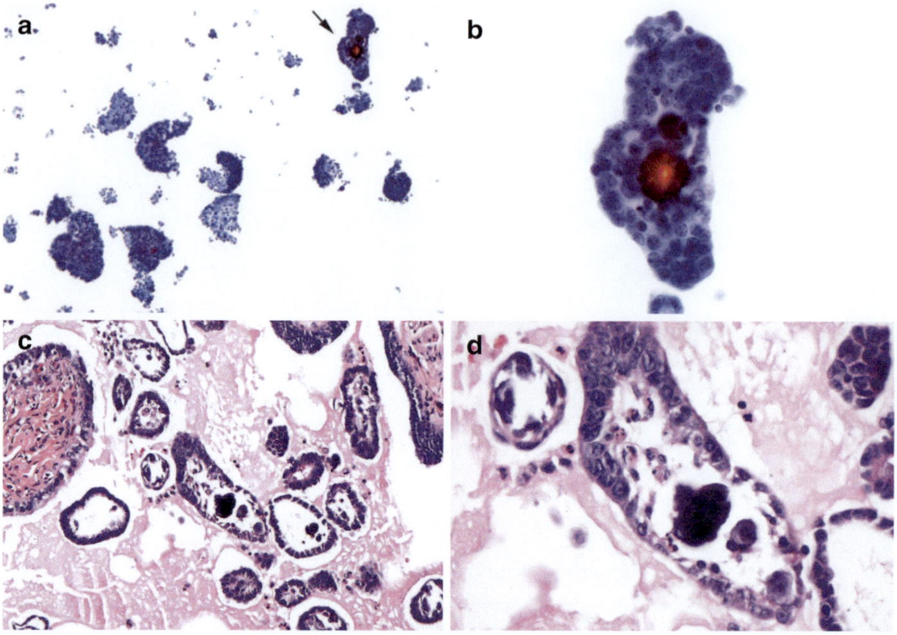

Fig. 5.9 Borderline serous tumor of the ovary. (**a**) The specimen has high cellularity with large hyperchromatic, papillary three-dimensional structures, smaller groups, and single cells. Note intact psammoma body in a group at 2 o'clock position (TP). (**b**) At closer examination, the three-dimensional group of tumor cells contain psammoma bodies. Note smooth "community" border, overlapping tumor cells with subtle cytological atypia. Peritoneal effusions may contain psammoma bodies in the absence of malignancy when mesothelial hyperplasia is present. Nuclear atypia is absent in benign mesothelial cells [6] (TP). (**c**, **d**) Histological section shows borderline tumor of the ovary. (**c**, H&E; **d**, H&E)

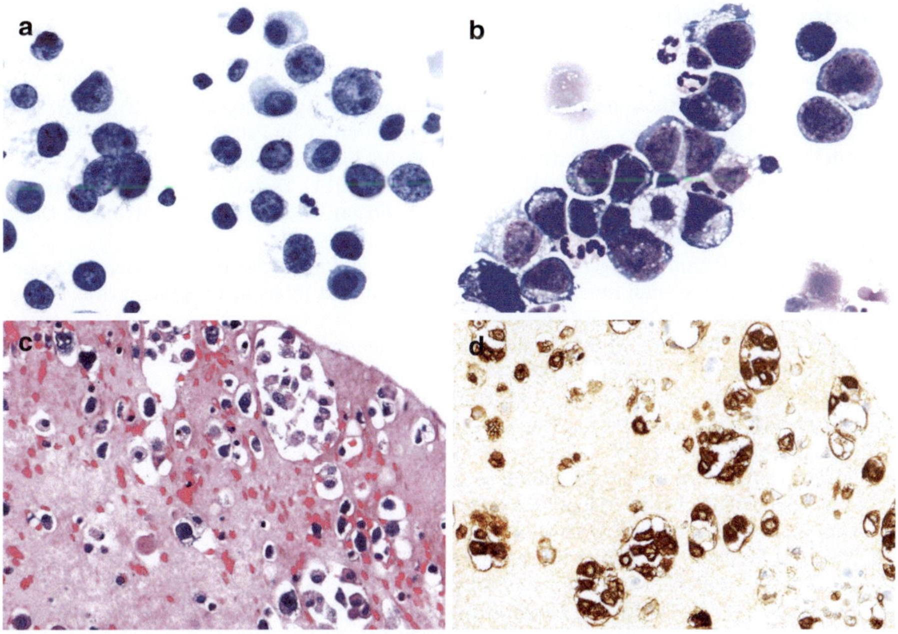

Fig. 5.10 Acute myeloblastic leukemia (AML). (**a**) Hematolymphoid processes can involve effusions and may blend with inflammatory cells. In this case, the patient has a history of AML which involved the pleural cavity. The cells are mostly singly dispersed, are large (compared to neutrophils), have delicate cytoplasm and enlarged nuclei with coarse chromatin on TP, and may be difficult to distinguish from mesothelial cells and other malignancies. Differential diagnosis with single malignant cell includes poorly differentiated carcinoma and melanoma (TP). (**b**) A Giemsa-based stain allows myeloid features of the malignant cells to be better recognized. The characterization of acute leukemias is based on a multiparametric analysis including clinical features, cell morphology, cytogenetics, and flow cytometry. The latter is important in identification of lineage. Additional material needs to be submitted for special studies (Giemsa-stained TP). (**c**) AML on cell block shows individual atypical cells, some binucleated, in lacunar spaces (H&E); (**d**) immunostain for CD163 was positive

Suggested Reading

1. Ylagan LR, Zhai J. The value of ThinPrep and cytospin preparation in pleural effusion cytological diagnosis of mesothelioma and adenocarcinoma. Diagn Cytopathol. 2005;32:137–44.
2. Nguyen G. Cytopathology of pleural mesotheliomas. Am J Clin Pathol. 2000;114:S68–71.
3. Lin O. Challenges in peritoneal cytology. Arch Pathol Lab Med. 2009;133:739–42.
4. Gong Y, Sun X, Michael CW, Attal S, Williamson BA, Bedrossian CW. Immunocytochemistry of serous effusion specimens: a comparison of ThinPrep vs cell block. Diagn Cytopathol. 2003;28:1–5.
5. Gamez RG, Jessurun J, Berger MJ, Pambuccian SE. Cytology of metastatic cervical squamous cell carcinoma in pleural fluid: report of a case confirmed by human papillomavirus typing. Diagn Cytopathol. 2009;37:381–7.
6. Parwani AV, Chan TY, Ali SZ. Significance of psammoma bodies in serous cavity fluid: a cytopathologic analysis. Cancer. 2004;102:87–91.
7. Gabriel C, Achten R, Drijkoningen M. Use of liquid-based cytology in serous fluids: a comparison with conventional cytopreparatory techniques. Acta Cytol. 2004;48:825–35.
8. Moriarty AT, Schwartz MR, Ducatman BS, et al. A liquid concept—do classic preparations of body cavity fluid perform differently than ThinPrep cases? Observations from the College of American pathologists interlaboratory comparison program in nongynecologic cytology. Arch Pathol Lab Med. 2008;132:1716–8.
9. Nicol TL, Kelly D, Reynolds L, Rosenthal DL. Comparison of TriPath thin-layer technology with conventional methods on nongynecologic specimens. Acta Cytol. 2000;44:567–75.
10. Nasuti JF, Tam D, Gupta PK. Diagnostic value of liquid-based (ThinPrep) preparations in nongynecologic cases. Diagn Cytopathol. 2001;24:137–41.
11. Elsheikh TM, Kirkpatrick JL, Wu HH. Comparison of ThinPrep and cytospin preparations in the evaluation of exfoliative cytology specimens. Cancer. 2006;108:144–9.
12. Hoda RS. Non-gynecologic cytology on liquid-based preparations: a morphologic review of facts and artifacts. Diagn Cytopathol. 2007;35:621–34.
13. Michael CW, McConnel J, Pecott J. Comparison of ThinPrep and TriPath PREP liquid-based preparations in nongynecologic specimens: a pilot study. Diagn. Cytopathol. 2001;25:177–84.
14. Leung CS, Chiu B, Bell V. Comparison of ThinPrep and conventional preparations: nongynecologic cytology evaluation. Diagn Cytopathol. 1997;16:368–71.
15. Fetsch PA, Simsir A, Brosky K, Abati A. Comparison of three commonly used cytologic preparations in effusion immunocytochemistry. Diagn Cytopathol. 2002;26:61–6.
16. Jing X, Li QK, Bedrossian U, et al. Morphologic and immunocytochemical performances of effusion cell blocks prepared using 3 different methods. Am J Clin Pathol. 2013;139:177–82.

Respiratory Exfoliative Cytology

6

Introduction

The respiratory tract is comprised of upper and lower airways. The upper airway extends from the sinonasal tract to the larynx, and the lower tract extends from the trachea to the lungs. Although all sites are amenable to cytological sampling, the lower tract is usually the target for the detection of infections, benign lesions, and neoplastic processes. Various sampling techniques are utilized, occasionally with the use of concurrent needle core biopsies (NCB). In this chapter only the lower respiratory tract will be addressed.

Lung carcinoma is common and is the leading cause of death in men and women. According to the American Cancer Society, there will be approximately 225,000 new cases of lung cancer and about 158,080 deaths from the disease (accounting for one of four cancer deaths) in 2016. Only 20 % of lung cancers are diagnosed at an early stage when the disease is still localized within the lungs. At the time of diagnosis, 25 % of patients have regional metastasis, and 55 % of patients have distant spread of disease. Thus, early detection remains the major cornerstone for the successful treatment of pulmonary malignancies.

Carcinomas, both of the small cell and non-small cell types, are by far the most common malignancy of the respiratory tract. Non-small cell lung cancer (NSCLC) accounts for ~85 % of all lung cancers. Histologically, NSCLC is divided into adenocarcinoma, squamous cell carcinoma, and large cell neuroendocrine carcinoma. Patients with NSCLC require a complete staging workup to evaluate the extent of disease because stage plays a major role in determining the choice of treatment.

The accuracy of differentiating between small cell carcinoma and NSCLC on cytology specimens ranges from 94 to 100 % when compared with resection or autopsy specimens. The accuracy of subclassifying NSCLC into adenocarcinoma, squamous cell carcinoma, and large cell neuroendocrine carcinoma ranges from 66 to 91 %. Concordance between bronchoscopically obtained cytology and biopsy specimens is >95 % in recent studies. The overall sensitivity increases with a combined use of different sampling modalities.

R.S. Hoda et al., *Diagnostic Liquid-Based Cytology*,
DOI 10.1007/978-3-662-53905-7_6, © Springer-Verlag GmbH Germany 2017

Metastatic malignant neoplasms, from almost any site, can metastasize to the lung. Generally, lung metastases are identified in 30–55 % of all cancer patients. Tumors that commonly metastasize to the lung include carcinomas from the colon, breast, prostate, and urinary bladder, sarcomas, and melanomas.

Cytological Reporting Guidelines

The new Papanicolaou Society of Cytology guidelines for standardized terminology and nomenclature for respiratory cytology are designed to stratify the risk of malignancy with diagnostic categories for appropriate patient management [1]. The current guidelines provide diagnostic categories and criteria and also describe techniques for obtaining specimens, ancillary testing, and patient follow-up and management. A six-tiered system is recommended as the standardized nomenclature for reporting respiratory cytology diagnoses. The categories proposed are nondiagnostic, negative (for malignancy), atypical, neoplastic (benign and low-grade malignancy), suspicious for malignancy, and malignant. A multidisciplinary diagnostic approach is recommended. Patient management should be determined by correlating the clinical findings in concert with imaging features, cytological findings, and result of molecular analysis, if pertinent.

Indications, Collection, and Laboratory Processing of Exfoliative Respiratory Tract Samples

The examination of exfoliative respiratory cytology is an efficient and cost-effective method for diagnosing respiratory tract lesions. An accurate diagnosis relies on receiving an adequate sample, optimal specimen preparation using LBP, cell block, and expertise in interpretation of LBP.

The principal indications for exfoliative cytology of the respiratory tract are as follows:

1. Workup of a solitary pulmonary nodule detected on diagnostic or screening CT
2. Workup of a pulmonary nodule detected on CT in patients with a prior malignancy
3. Workup of pulmonary infiltrates to exclude infectious etiology

Methods of Collection for Exfoliative Cytological Samples

- Sputum
- Bronchial brush and bronchial wash
- Bronchoalveolar lavage (BAL)

Sputum

The cytological examination of a spontaneously expectorated sputum is a noninvasive technique and can be used for the detection of infectious lesions and large centrally located tumors. Diagnostic sensitivity of sputum cytology is directly proportional to the number of samples examined, and its accuracy approaches 95 % when five (5) sequential samples are examined. However, sputum cytology is not recommended for screening lung cancer. A post-bronchoscopic sputum may be examined when the expectorated sputum is negative.

Bronchial Brush and Bronchial Wash Specimens

Bronchial brushing and bronchial wash specimens are obtained via flexible fiberoptic bronchoscope (FOB). FOB can also obtain fine-needle aspiration (FNA) biopsies and transbronchial forceps biopsies. The advantages of FOB include an increased visual range, especially in the upper lobes, minimal discomfort to the patient, and sampling of previously inaccessible lesions such as peripheral nodules.

Both bronchial brushing and bronchial wash can detect infectious and neoplastic lesions. If indicated, microbiological cultures can also be performed concurrently. The diagnostic yield for bronchial brushing depends upon an adequate bronchoscopic sampling as well as the size and location of the lesion. For central tumors, the diagnostic sensitivity of bronchial brushing and bronchial wash is lower than that of transbronchial biopsy (TBBx). The sensitivity of bronchial brushing and bronchial wash is further decreased for peripheral lesions. The concordance rate of bronchial brushing and TBBx is ~97 %. The highest diagnostic yield for bronchial brushing and bronchial wash is for squamous cell carcinoma, adenocarcinoma, and small cell carcinoma (in that order).

Bronchoalveolar Lavage (BAL) Specimens

BAL also performed using FOB is a safe technique with a diagnostic accuracy comparable to TBBx. It is valuable for the detection of opportunistic infections, and since it samples multiple bronchi, it is also suitable for sampling diffuse lesions such as adenocarcinoma with a lepidic growth pattern. The diagnostic sensitivity of BAL for organisms in immunosuppressed people is 82 % for *Pneumocystis jirovecii*, 83 % for cytomegalovirus (CMV) and fungal pneumonia, and 80 % for mycobacterial disease. The overall reported diagnostic yield of BAL for malignancy is about 50 % with a lower yield for peripheral lesions. The diagnostic yield is improved with the addition of other modalities such as bronchial brushing, bronchial wash, and TBBx.

Laboratory Processing of LBP Specimens

Exfoliative respiratory tract specimens are usually processed as one Papanicolaou (Pap)-stained LBP. The collection specimen is rinsed in a preservative medium for LBP. Direct smears from brushings can also be made with a quick rolling motion of the brush on glass slides. The slides can be fixed in 95 % ethanol for Pap staining or air-dried for Romanowsky staining (e.g., Diff-Quik stain). Residual material can be rinsed in collection medium and used to process additional LBP for special stains, immunostains, or cell block preparation, especially if tissue fragments are noted. Please see Chap. 1 also.

Advantages of Cytological Specimens for Respiratory Tract over Needle Core Biopsy (NCB)

- Allows sampling of larger area of concern
- Allows sampling of narrow areas by brushings
- Shorter turnaround time
- Provides high quality of DNA for molecular testing

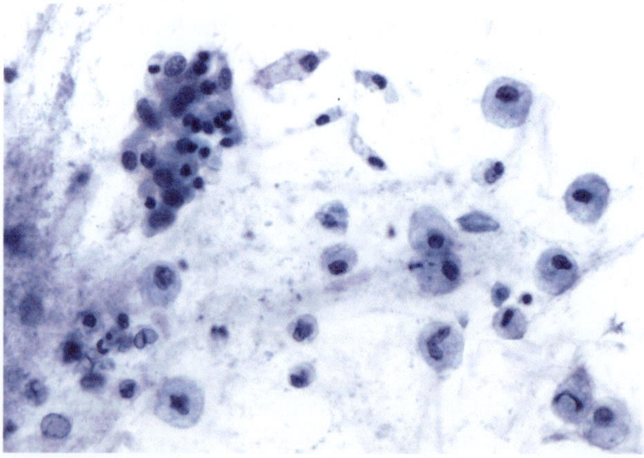

Fig. 6.1 Benign bronchial wash. This field shows a mucoid background with embedded benign bronchial cells and scattered macrophages. Note the well-preserved benign bronchial cells which are uniform with minimal overlapping. Terminal bar and cilia are visible in some cells. The macrophages retain finely vacuolated cytoplasm and pale kidney-shaped or oval nuclei. The mucin stains a purplish color and is of moderate thickness (TP). Excessive mucus, as well as blood and inflammatory cells, can be problematic in the processing and screening of LBP [2]. TP may show a loss of cellularity with large areas of the filters showing nearly complete absence of cells. This appears to be the result of excess mucin covering or obstructing the filtration membrane which diminishes cell retrieval, potentially impacting the detection of disease. In SP, cell enrichment process better manages excess mucus and does not affect cell recovery

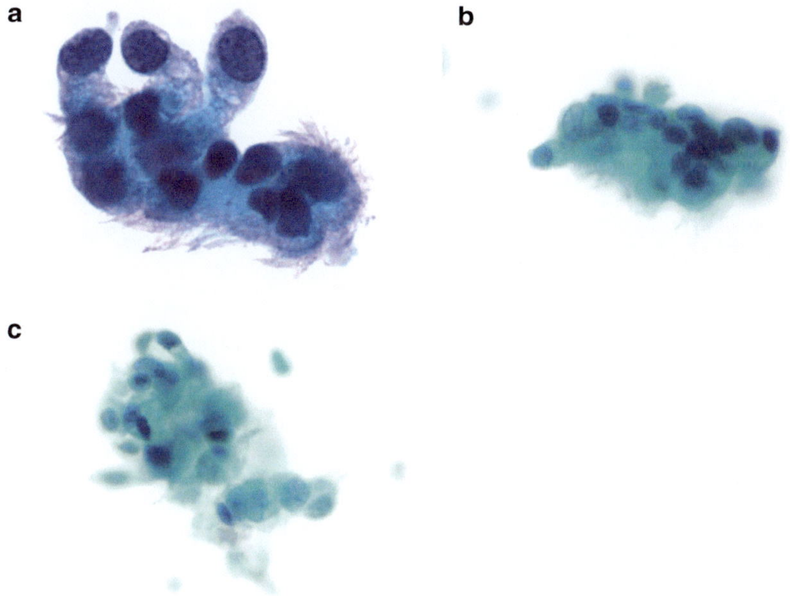

Fig. 6.2 Benign bronchial cells from bronchial wash. (**a**) The cells are elongated with oval apically placed nuclei with condensed chromatin with cytoplasmic terminal bars and eosinophilic-staining cilia (TP). (**b**) Benign bronchial cells in SP show features similar to TP (SP). (**c**) In reactive conditions, benign bronchial cells with cilia may form clusters with relatively high nuclear-to-cytoplasmic ratio and subtle nuclear atypia. Such reactive groups are termed "Creola bodies" and are in the differential diagnosis of well-differentiated adenocarcinoma. Cilia are the most salient distinguishing feature (SP)

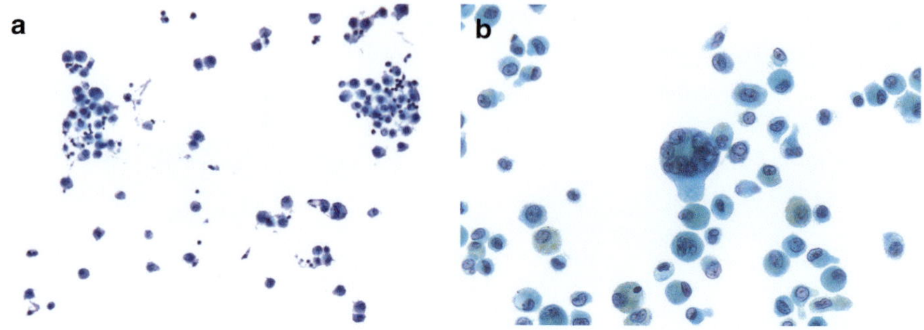

Fig. 6.3 Benign bronchoalveolar lavage. (**a**) An adequate BAL specimen must sample the alveolar spaces, as determined by the presence of pulmonary alveolar macrophages, as seen here intermixed with lymphocytes and small amounts of mucin. In LBP, the background is clean. Note the pulmonary alveolar macrophages clustering together and found singly with mucin attached to cell clusters (TP). (**b**) Pulmonary alveolar macrophages may contain hemosiderin pigment and have greenish-yellow staining cytoplasm on Pap stain. Binucleated and multinucleated cells are common and not suggestive of any specific disease process. The cells have abundant, foamy cytoplasm with an eccentrically placed, hyphen-shaped nucleus. Small nucleoli and chromocenters may be seen (TP)

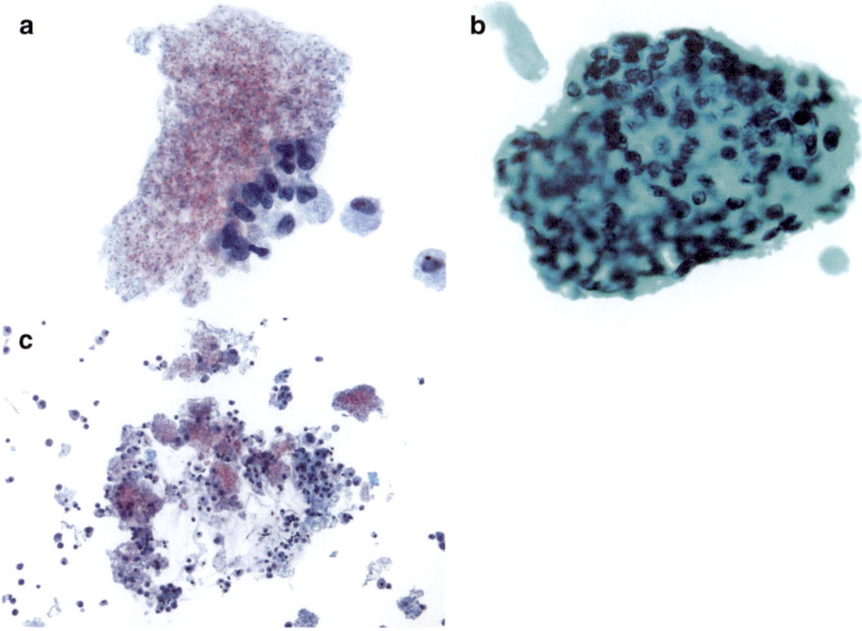

Fig. 6.4 Bronchoalveolar lavage with *Pneumocystis jirovecii*. (**a**) This specimen taken from a 40-year-old HIV-positive patient shows the characteristic foamy alveolar cast of *Pneumocystis* organisms. The casts have a bubble-like appearance and comprise of clusters of organisms that are spherical to ovoid with a smooth contour on Papanicolaou stain. The central eosinophilic and peripheral basophilic staining and trophozoites appearing as microdots within the organism's shell are characteristics for *Pneumocystis*. The organisms are also readily identified on Giemsa, Wright, Gram, and H&E stains. Ultrastructurally, the frothy appearance to the cast is due to filopodia connecting the organisms (TP). (**b**) Cup-shaped *Pneumocystis jirovecii* cysts (Grocott methenamine silver [GMS] stain). GMS may be helpful in distinguishing various fungi. *Pneumocystis* is more likely found in bronchial wash specimens. The latter sample has distal bronchial and alveolar spaces and has a diagnostic yield of 82–94 %. The background in BAL specimens with *Pneumocystis* is typically clean. Alveolar proteinosis is a potential mimic of *Pneumocystis jirovecii*. The central eosinophilic and peripheral basophilic staining and trophozoite microdots of *Pneumocystis* are lacking in alveolar proteinosis (Fig. 6.5). (**c**) Numerous intact alveolar casts of *Pneumocystis* organisms are present despite the multiple vigorous steps involved in TP processing (TP)

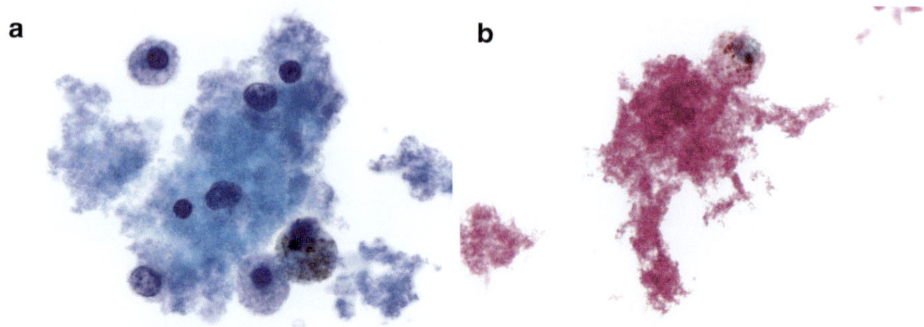

Fig. 6.5 Bronchoalveolar lavage with pulmonary alveolar proteinosis. (**a**) This entity is in the differential diagnosis of *Pneumocystis* organisms. A frothy, bubbly alveolar cast is seen surrounded by alveolar macrophages, some containing a suggestion of proteinaceous material and hemosiderin. No epithelial cells are noted. Unlike *Pneumocystis*, neither the ghosts of organisms nor microdots are visible. The background is clean (TP). (**b**) Special stains performed on additional TP slides showed the material to be diastase-resistant periodic acid-Schiff (PAS-D) positive, but negative for Alcian blue, GMS, and Congo red (PAS-D stain on TP). Pulmonary alveolar proteinosis is a rare disease characterized by abnormal intra-alveolar accumulation of extracellular surfactant-like material, which is composed of proteins and lipids [3]. Grossly, BAL may be somewhat milky. Electron microscopy shows type II pneumocytes containing concentrically laminated structures, some of which contain dense osmophilic cores, amid background proteinaceous debris

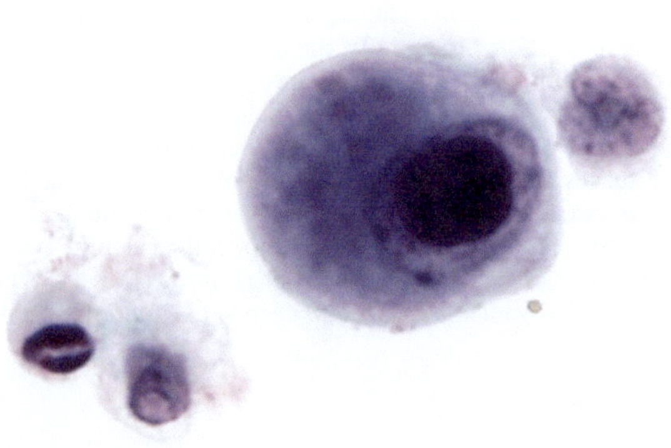

Fig. 6.6 Bronchoalveolar lavage with cytomegalovirus (CMV). CMV pneumonitis in an immunocompetent patient shows the virus in a glandular cell with cellular enlargement (compare its size with that of the inflammatory cells), nuclear enlargement, and large intranuclear basophilic inclusion surrounded by a halo ("owl's eye") (TP). Occasionally, smaller nuclear or cytoplasmic inclusions may also be seen. Multinucleation is uncommon in CMV-infected cells. BAL is a major diagnostic tool in lung diseases, including the detection of viral respiratory infections with herpes simplex virus (HSV), CMV, and Epstein-Barr virus (EBV) among others. Rapid detection of CMV pneumonitis is reliable using PCR testing of BAL cells. Immunocytochemistry and histologic evaluation of lung parenchyma obtained by TBBx may also be used to increase detection

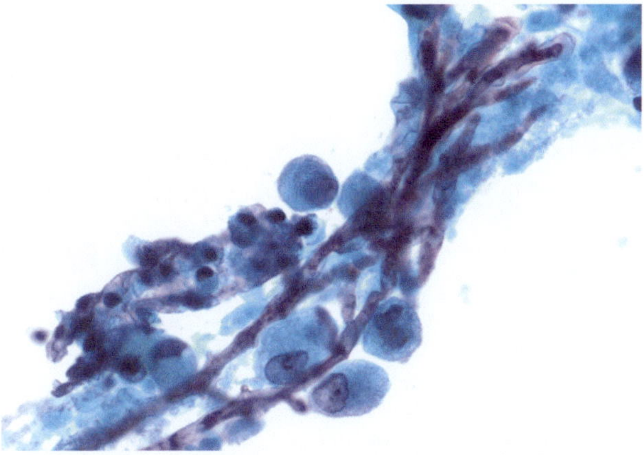

Fig. 6.7 Bronchial wash with aspergillus. The field shows septate hyphae, 3–6 μm in size and with a 45° angle branching pattern (TP). The background shows mucus and macrophages. Aspergillus is a common fungus and is transmitted by airborne conidia. It can occur in immunocompetent and immunosuppressed individuals. *Aspergillus fumigatus* is the most frequent human pathogen. The fungi form conidial head or "fruiting bodies" that produce spores. These spores can easily be airborne. *Aspergillus spp.* cannot be morphologically distinguished from its mimics such as Zygomycetes and *Candida spp.* unless it is accompanied by fruiting bodies. The latter are important in the identification of particular *Aspergillus* species and in differentiating *Aspergillus* from its mimics

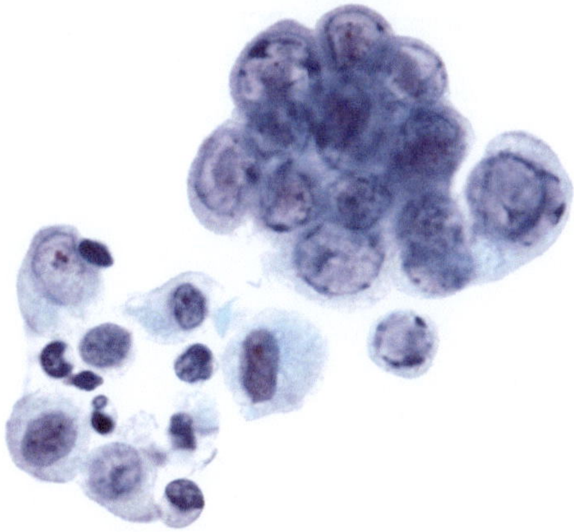

Fig. 6.8 Bronchoalveolar lavage with herpes simplex virus: HSV-infected cells have enlarged nuclei with a homogenous "ground glass" appearance. The nuclei mold against one another, and individual cells may be multinucleated. Note also the dark condensation seen along the nuclear rim which contrasts with the pale nucleus. Inflammatory cells are present adjacent to the infected cells (TP)

a
b
c

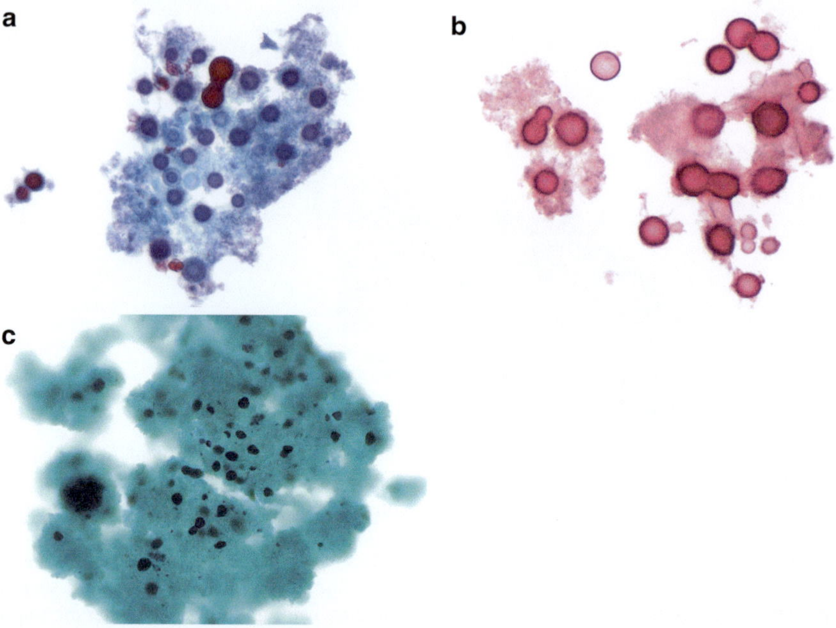

Fig. 6.9 Bronchoalveolar lavage with cryptococcus. (**a**) A single spherical organism surrounded by a thick capsule is noted with pulmonary macrophages (TP). (**b**, **c**) Narrow-based budding is prominent on mucin (mucicarmine) stain and GMS; the former stains the organisms bright red and the latter black. Note the translucent capsule seen on mucicarmine stain (**b**, mucicarmine stain on TP, C, GMS stain on TP). A multiplex real-time PCR technique has been developed for the detection of the three most common causes of fungal opportunistic pneumonia in AIDS patients: *Pneumocystis*, *Histoplasma capsulatum*, and *Cryptococcus*

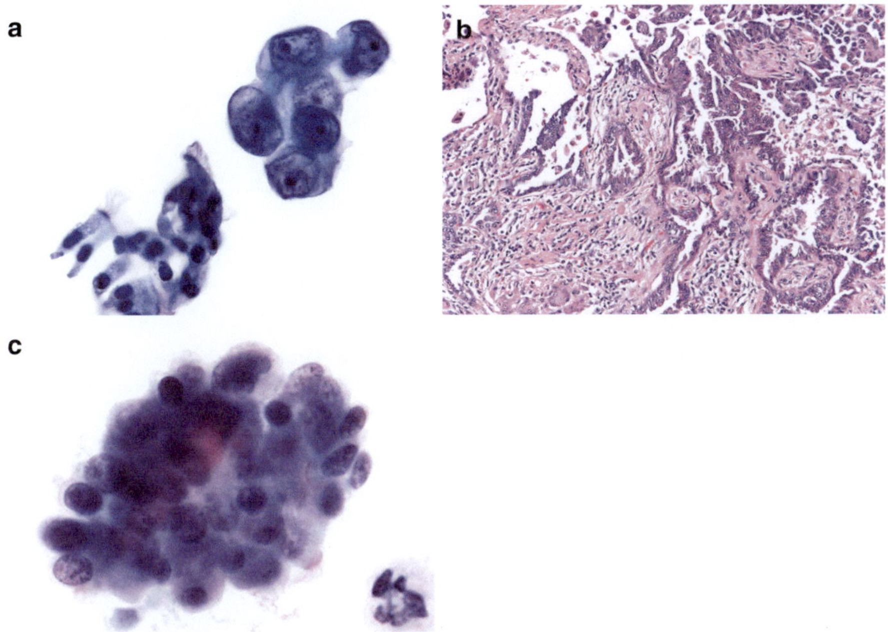

Fig. 6.10 Bronchial wash with adenocarcinoma. (**a**) Cluster of malignant cells is evident. The nuclear features of malignancy can be appreciated, with enlarged nuclei with irregular borders, thick membranes and macronucleoli. The nuclear-to-cytoplasmic ratio is high. Compare these cells to benign respiratory epithelial cells seen in the bottom left. Cluster formations are retained in TP with better preservation of their nuclear features as compared to conventional smear (TP). (**b**) TBBx shows adenocarcinoma. Tumor cells were immunoreactive for TTF-1 and Napsin-A, both indicative of lung origin (H&E). (**c**) Creola body on TP. Compare with SP in Fig. 6.2c. Note cilia faintly visible in some cells and terminal bars without distinct cilia in other cells (TP)

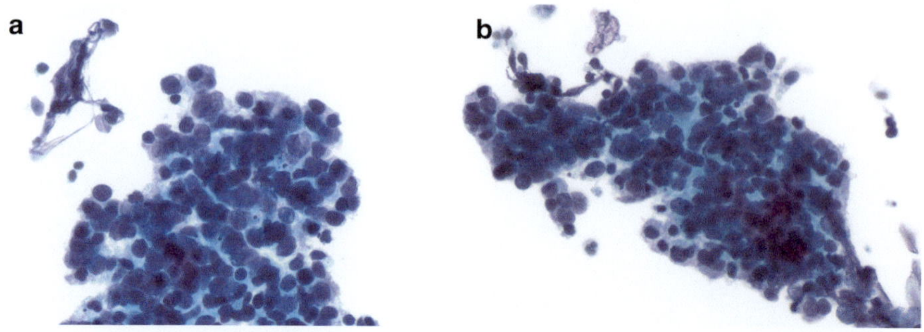

Fig. 6.11 Bronchial wash with small cell carcinoma. **(a, b)** Group of small cell carcinoma cells with high nuclear-to-cytoplasmic ratios and apoptotic bodies. Nuclei show molding, irregularity, condensed and coarse chromatin, occasional small nucleoli, and minimally discernible cytoplasm (TP). Background shows crush artifact. Some granular diathesis clings to tumor cells. Kim et al. [4] examined a series of small cell carcinoma on TP and found that the cells could demonstrate shrinkage artifact and display small nucleoli and nuclei with odd shapes (dumbbell and pencillike). Nuclear molding and nuclear streaks were rarely seen; but apoptotic cells and granular necrotic debris could be faintly discerned in the background. In small cell carcinoma, nuclear chromatin details were better observed on TP; however, there was loss of spindling and nuclear molding observed in direct smears

Suggested Readings

1. Layfield LJ, Baloch Z, Elsheikh T. Standardized terminology and nomenclature for respiratory cytology: the Papanicolaou society of cytopathology guidelines. Diagn Cytopathol. 2016;44:399–409.
2. Kenyon S, Sweeney BJ, Happel J. Comparison of BD Surepath and ThinPrep Pap systems in the processing of mucus-rich specimens. Cancer Cytopathol. 2010;118:244–9.
3. Narine N, Rana DN, Santhanakrishnan K. Cytological and electron microscopic findings in a bronchoalveolar lavage sample from a case of pulmonary alveolar proteinosis with radiological correlation. Cytopathology. 2015; doi:10.1111/cyt.12301.
4. Kim S, Owens CL. Analysis of ThinPrep cytology in establishing the diagnosis of small cell carcinoma of lung. Cancer. 2009;117:51–6.
5. Tyler KL, Selvaggi SM, Stewart J 3rd. *Pneumocystis jirovecii* in a bronchoalveolar lavage specimen on ThinPrep® cytology. Diagn Cytopathol. 2011;39:675–6.
6. Rana DN, O'Donnell M, Malkin A. A comparative study: conventional preparation and ThinPrep 2000 in respiratory cytology. Cytopathology. 2001;12:390–8.
7. Choi YD, Han CW, Kim JH. Effectiveness of sputum cytology using ThinPrep method for evaluation of lung cancer. Diagn Cytopathol. 2008;36:167–71.
8. Elsheikh TM, Kirkpatrick JL, Wu HH. Comparison of ThinPrep and cytospin preparations in the evaluation of exfoliative cytology specimens. Cancer. 2006;108:144–9.
9. Hoda RS, Colello C, Roddy M. "Fruiting body" of Aspergillus species in a routine cervicovaginal smear (Pap test). Diagn Cytopathol. 2005;33:244–5.
10. Collins GR, Thomas J, Joshi N. The diagnostic value of cell block as an adjunct to liquid-based cytology of bronchial washing specimens in diagnosis and subclassification of pulmonary neoplasms. Cancer Cytopathol. 2012;120:134–41.
11. Choi YD, Han CW, Kim JH. Effectiveness of sputum cytology using ThinPrep method for evaluation of lung cancer. Diagn Cytopathol. 2008;36:167–71.

Fine Needle Aspiration of Thyroid Gland

7

Introduction

Fine needle aspiration (FNA) is used to triage clinically and radiologically detected thyroid nodules to a specific management, surgery (either total- or hemithyroidectomy), or surveillance. Needle core biopsies of the thyroid are not usually performed, and thus the FNA diagnosis often decides the need for surgery. The most common malignancy of the thyroid is papillary thyroid carcinoma (PTC), and due to its multifocal nature, a diagnosis of malignancy on FNA typically results in a total thyroidectomy. A benign diagnosis is reassuring and allows the patient to avoid surgery and undergo surveillance. Most laboratories report thyroid FNA diagnoses according to "The Bethesda System for Reporting Thyroid Cytology (TBSRTC)," which contains six categories, each with their own rate of malignancy and recommended clinical management (Table 7.1) [1]. Molecular tests are now commonly performed on indeterminate thyroid nodules to help triage patient treatment.

Rapid on-site evaluation (ROSE) is commonly performed during thyroid FNA, in which DS are made from the FNA material on-site and stained with Diff-Quik, or a similar stain, to allow for rapid microscopic examination. In such instances, needle rinses may be saved for LBP, allowing for additional material to be visualized. However, many endocrinology offices perform thyroid FNA without ROSE and make multiple DS to insure adequacy of the sample. Nuclear features are important in the diagnosis of PTC, and thus most pathologists prefer to examine specimens using the Pap stain. However, such material must not be air-dried and immediately fixed in alcohol prior to drying. Clinicians often do not appreciate how quickly the air-drying artifact occurs and do not see the impact of poor preparation. In a busy office, the clinician may unintentionally allow the slides to air-dry prior to placement in alcohol. This results in a suboptimal or even nondiagnostic specimen. In order to improve the quality of thyroid specimens and reduce the number of slides per specimen, needle passes can be placed directly and entirely into LBP fixation solution and sent to the laboratory for processing using TP or SP. This allows for only one Pap-stained slide to be created for screening and greatly minimizes the

R.S. Hoda et al., *Diagnostic Liquid-Based Cytology*,
DOI 10.1007/978-3-662-53905-7_7, © Springer-Verlag GmbH Germany 2017

Table 7.1 TBSRTC (Adapted from [1])

Category	Rate of malignancy (%)	Usual management
Nondiagnostic	N/A	Repeat FNA
Benign	0–3	Surveillance
Atypia of undetermined significance (AUS) or follicular lesion of undetermined significance (FLUS)	5–15	Repeat FNA
Follicular neoplasm (FN) or suspicious for a follicular neoplasm (SFN)	15–30	Lobectomy
Suspicious for malignancy (SFM)	60–76	Lobectomy or near-total thyroidectomy
Malignant	97–99	Near-total thyroidectomy

chances of the specimen air-drying. In addition, it minimizes the work required on the part of the clinician performing the FNA, as no glass slides have to be prepared. Needle passes can be taken in quick succession, minimizing the time of the procedure for both the clinician and the patient.

Studies have shown that one TP slide performs equally to multiple DS made from one nodule. TP may perform superiorly in some circumstances [2]. However, other studies have demonstrated that LBP may have some limitations when compared to CS. For instance, one study has demonstrated that background lymphocytes may not be seen as prominently on LBP than on DS, resulting in decreased identification of Hashimoto's thyroiditis (HT) on LBP [3].

Familiarity with the cytomorphological appearance on LBP is critical for arriving at the proper diagnosis. Studies have shown that LBP causes certain artifacts and thus LBP morphology that sometimes differ from that seen on DS. Most importantly, thin colloid often has a delicate "tissue paper-type" appearance that is not seen on DS [4]. Thyroid follicular cells often shrink and appear smaller than what is seen on Pap-stained DS. Other differences have been described elsewhere in the literature [5].

The use of ancillary molecular tests in conjunction with indeterminate diagnoses is often used to further guide patient management. The commercially available molecular tests include Afirma (VeraCyte Inc., San Francisco, CA) and ThyroSeq (UPMC, Pittsburg, PA). Details of these tests have extensively been published and recently reviewed [6]. One benefit of LBP is that residual sample is available to be used for molecular testing. A recent study has shown that good-quality RNA can be extracted from residual liquid-based samples and used for the detection of BRAF and RAS point mutations as well as RET/PTC and PAX8/PPARϒ rearrangements. This avoids obtaining additional passes or having the patient come back for repeat biopsy, if the FNA diagnosis is indeterminate [7]. Molecular test can also be performed on the same diagnostic liquid-based slide. DNA has been shown to remain stable at room temperature over a 1-year period of storage in LBP media. RNA, however, may rapidly degrade at room temperature and needs freezing at −80 °C to maintain integrity [8]. Studies have shown that molecular testing for thyroid nodules with indeterminate cytology improves presurgical diagnoses and significantly reduces unnecessary thyroid surgeries [9].

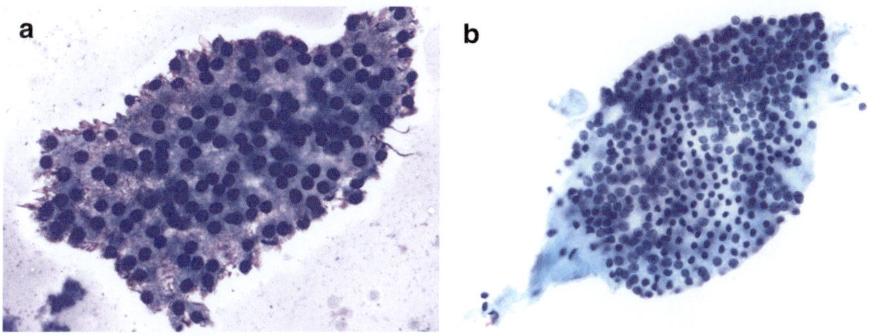

Fig. 7.1 Benign follicular cells. (**a**) Sheet of benign follicular cells with evenly spaced regular round and uniform nuclei, abundant lacy cytoplasm, and low nucleus-to-cytoplasmic ratio. Thin colloid appears diffusely spread around the sheet of cells (DQ-stained DS, HP). (**b**) Similar features can be seen in this LBP, except the background is clean. Colloid adheres to the follicular cell group. Note the smooth contour of the group. The Pap-stained LBP allows better assessment of the nuclear detail, in this case revealing bland chromatin with tiny nucleoli (TP)

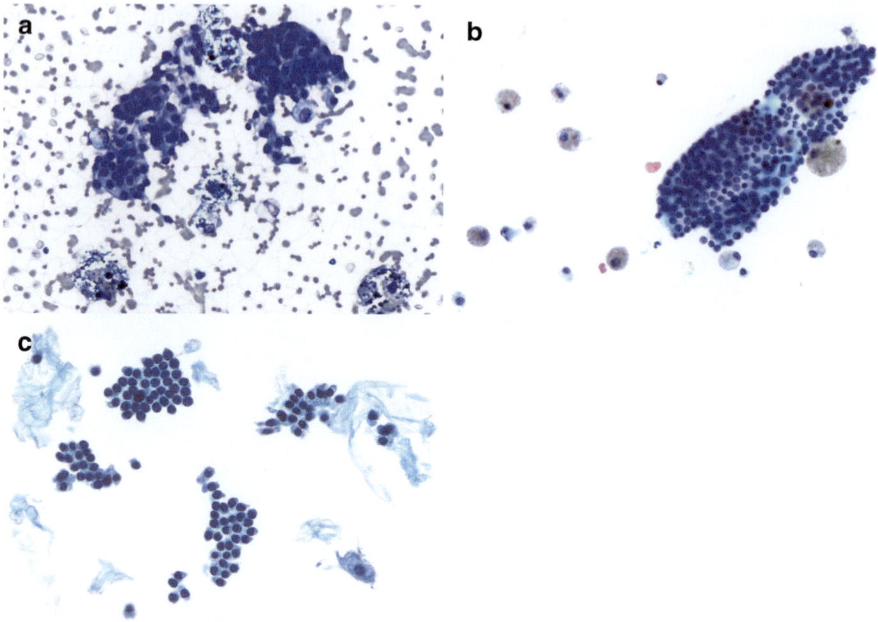

Fig. 7.2 Nodular hyperplasia. (**a**) A mixture of benign follicular cell fragment, diffuse thin watery colloid, with the appearance of "cracked desert sand," and hemosiderin-laden macrophages are present. The pigment in DQ stain appears green-blue. Here the follicular cells have a similar appearance to that described in Fig. 7.1a (DQ-stained DS); (**b**) on LBP, the background is clean and thin colloid appears as small clumps of delicate fragments known as "tissue paper" colloid [4]. The pigment in cystic macrophages has a golden-brown appearance on the Pap stain. Most importantly, the follicular cells have a benign appearance as seen in Fig. 7.1b (TP). (**c**) Big and small flat sheets of uniform follicular cells with less cytoplasm but the nuclei do not overlap and are evenly spaced; in other fragments the cells have a columnar appearance. Note the abundant thin delicate colloid in the background. The flat sheet of organized and uniform follicular cells and abundant thin colloid strongly suggest a benign nodule (TP)

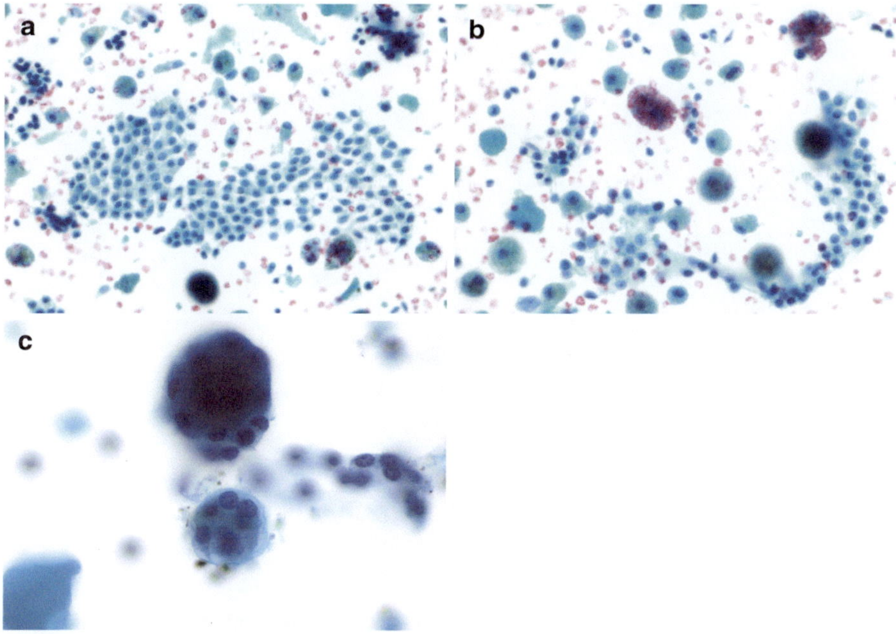

Fig. 7.3 Nodular hyperplasia. (**a**) The appearance of nodular hyperplasia on SP preparation is similar to that of TP. Note the large and small flat sheets of evenly spaced follicular cells and crisp nuclear morphology. Thin colloid appears clumped and many hemosiderin-laden macrophages are evident (SP). (**b**) In this image balls of colloid appear deeply eosinophilic in the center. Ball of colloid on the left side appears dense and basophilic. The staining color of colloid may indicate its thickness, with dense colloid staining in a two-tone color, eosinophilic inside and basophilic on the outer aspect. (**c**) Note the three-dimensional cells at high magnification forming microfollicles from the same case, which can occasionally be seen with adenomatoid change in nodular hyperplasia (SP). It is important to become familiar with the various appearances of colloid on LBP

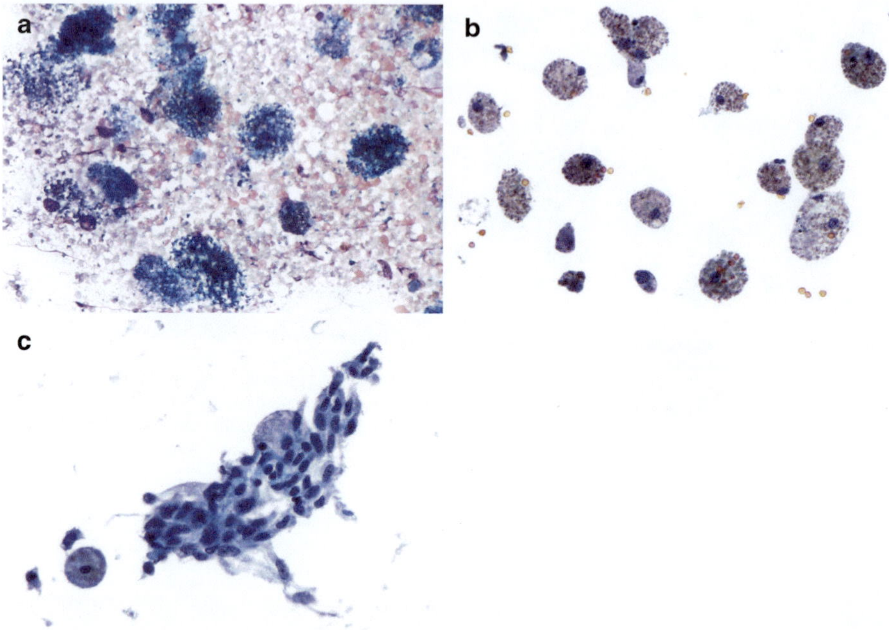

Fig. 7.4 Cyst fluid. (**a**) The aspiration of some cystic lesions may result in cyst fluid without sampling of any follicular lining cells. The fluid contains numerous hemosiderin-laden macrophages, with pigment that appears green, brown, and blue in the cytoplasm – this is the result of the degradation of red blood cells over time. The background also contains red cells, degenerated blood, and debris. Colloid is very thin and appears as small linear bluish-purple streaks and also indicated by the empty spaces between the macrophages. Thin colloid may be lost if the slide is moved vigorously during DQ staining (DQ-stained DS). (**b**) The TP from the same case has a clean background, creating a stark contrast to the DQ-stained smear. The field contains hemosiderin-laden macrophages and scant clumped thin colloid on the left lower aspect, indicating the presence of cyst fluid (TP). (**c**) When cyst-lining cells are sampled, they may have an unusual appearance and seem atypical. The cells often have elongated "stretched" cytoplasm and mildly atypical oval to spindled nuclei (TP). Faquin et al. [10] observed that atypical cells from benign thyroid cysts appear elongated and form small cohesive flat sheets that display a streaming appearance with defined cell borders and windows reminiscent of repair. The cells may have pulled-out, spindled morphology with intranuclear grooves, pale chromatin, and granular cytoplasm

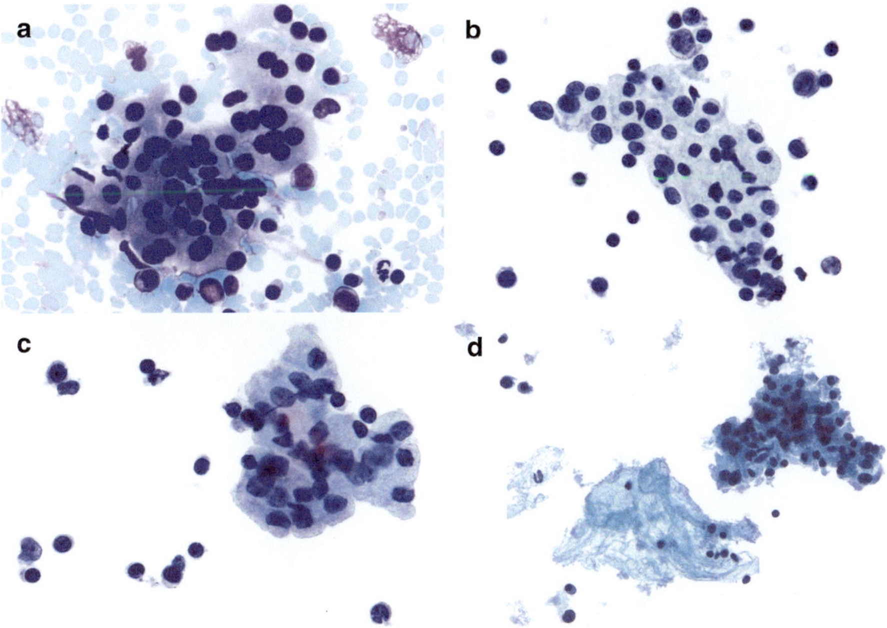

Fig. 7.5 Hashimoto thyroiditis. (**a**) Polymorphous lymphocytes and Hurthle cells comprise Hashimoto thyroiditis. Hurthle cells are seen as large cells with single nucleoli and moderately abundant eosinophilic granular cytoplasm. Bi- or multinucleated Hurthle cells are also present. Lymphocytes are in proximity to Hurthle cells and are benign and polymorphous (DQ-stained DS). (**b**, **c**) The background is clean except for heterogeneous population of lymphocytes associated with the Hurthle cells. The Hurthle cells have round and regular nuclei with nucleoli; anisonucleosis is a common feature of Hurthle cells in Hashimoto thyroiditis. Cytoplasm may appear soft and granular or dense (TP). (**d**) Hashimoto thyroiditis shows similar findings in SP specimen except for more evident colloid (SP). On LBP, there may be fewer mononuclear cells, or they may occur mostly in aggregate or toward the periphery of the slide [3]

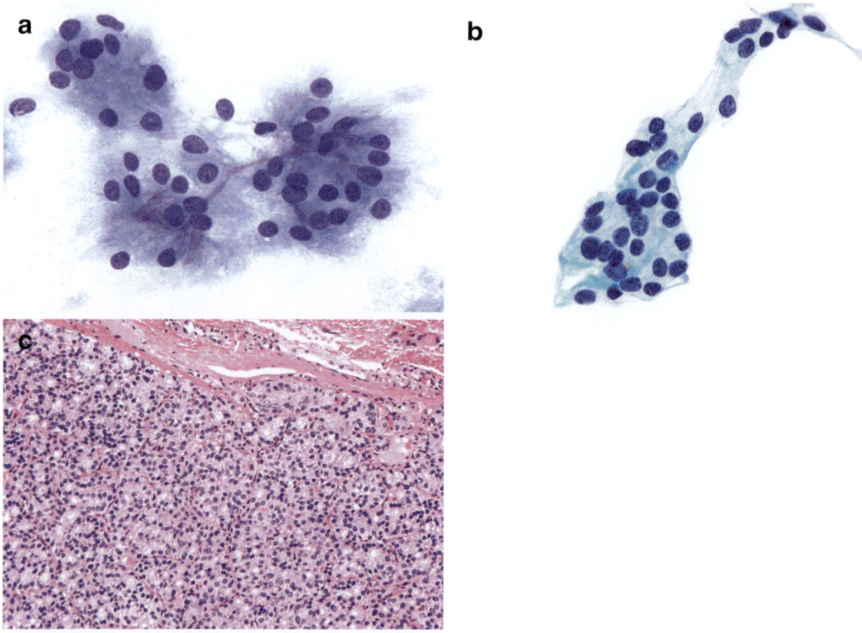

Fig. 7.6 Suspicious for a follicular neoplasm. (**a**) This diagnosis is used when a cellular specimen contains thyroid follicular cells in a predominantly microfollicular architecture – either as individual microfollicles or larger groups of microfollicles. The follicular cells in this image form small "rings" with less than ten cells present in each microfollicle "ring." The cells are otherwise bland, with round to oval nuclei with regular borders, nucleoli, and only minimal nuclear overlap. Intranuclear grooves are not present and nucleus-to-cytoplasmic ratio is low (DQ-stained DS). (**b**) A trabecula of microfollicles associated with moderately thick colloid. Nuclear morphology is well preserved (TP). (**c**) Histologic diagnosis was follicular adenoma. Compare the cytological microfollicles with its histologic correlate (H&E). Histologic differential diagnosis of a cytological diagnosis of "suspicious for a follicular neoplasm" includes adenomatoid nodules in a nodular hyperplasia, follicular adenomas, follicular carcinomas, or follicular variants of PTC

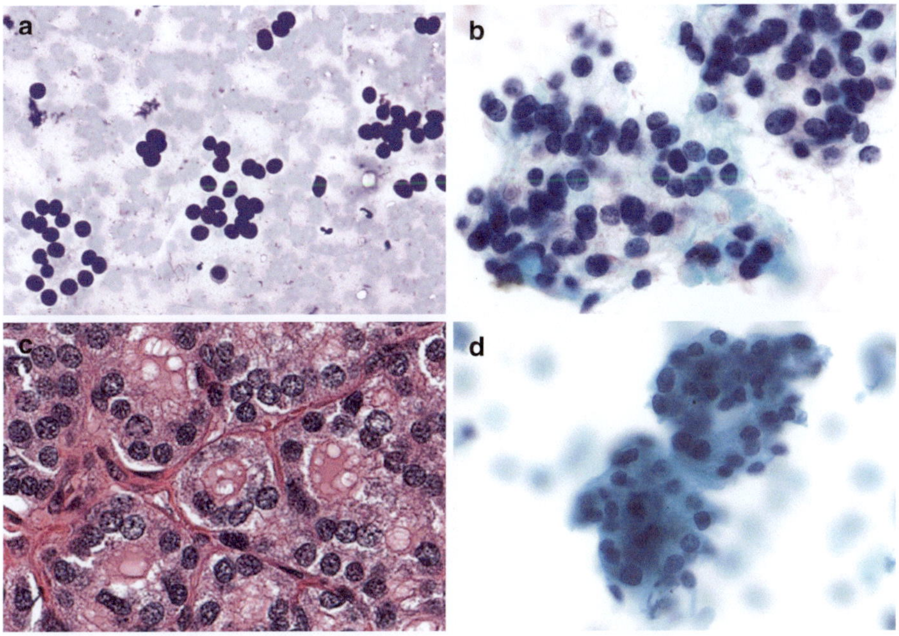

Fig. 7.7 Follicular neoplasm. (**a**) FA are histologically distinguished from follicular carcinomas by the absence of capsular or vascular invasion, which is not a feature that can be determined cytologically. In this case of FA, FNA reveals a moderately cellular smear with follicular cells in distinct and individually dispersed microfollicles. The follicular cells appear devoid of cytoplasm. Nuclei are slightly enlarged (compared with RBCs), bland appearing, round, and regular. Nucleoli are not evident. Nuclear features of PTC are not present. Colloid is extremely scant and visible as small purplish foci. This was diagnosed as "suspicious for a follicular neoplasm," based on microfollicular pattern (DQ-stained DS). (**b**) The smear is cellular with distinct microfollicles. Nuclear features of enlargement, nucleoli, overlap and crowding and hyperchromasia are evident. Cytoplasm is extremely scant and some thick colloid is present close to the cells (Pap-stained DS). (**c**) Histologic diagnosis was follicular adenoma. Note microfollicular architecture and monotonous, round, and regular nuclei (H&E). (**d**) SP shows a conglomerate of microfollicles. Cytoplasm is granular and dense and nuclear features are similar to DS and TP. Since the microfollicles have a three-dimensional structure, it makes the background appear out of focus (SP)

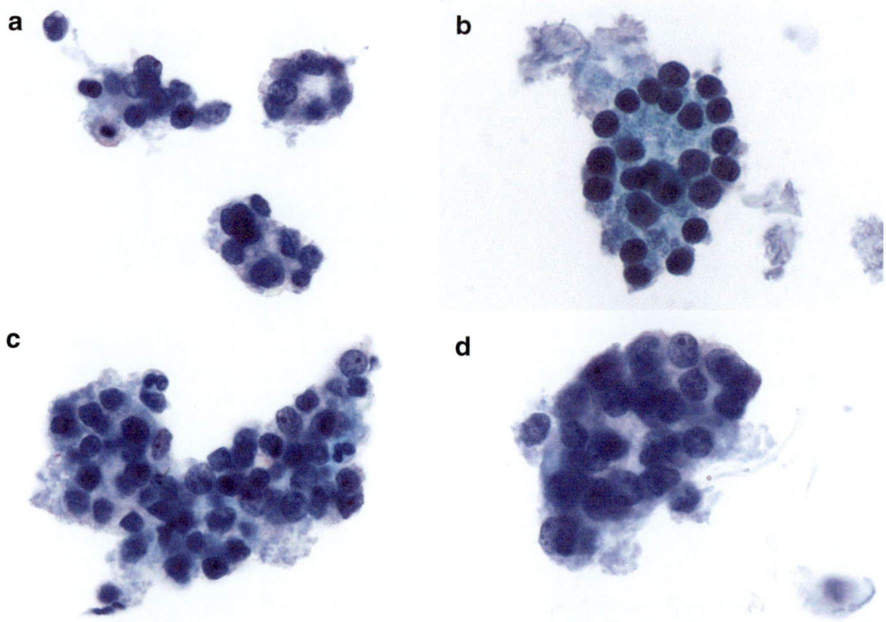

Fig. 7.8 Follicular neoplasm. (**a–d**) Follicular cells in a microfollicular architecture, present singly or in a fragment. Colloid is closely associated with the cells. Nuclei are atypical and are enlarged, overlapping, and overcrowded, with membrane irregularity, hyperchromasia, nucleoli, and occasional grooves. Cytological diagnosis was "follicular neoplasm," and a subsequent lobectomy revealed a follicular carcinoma showing capsular and vascular invasion (TP)

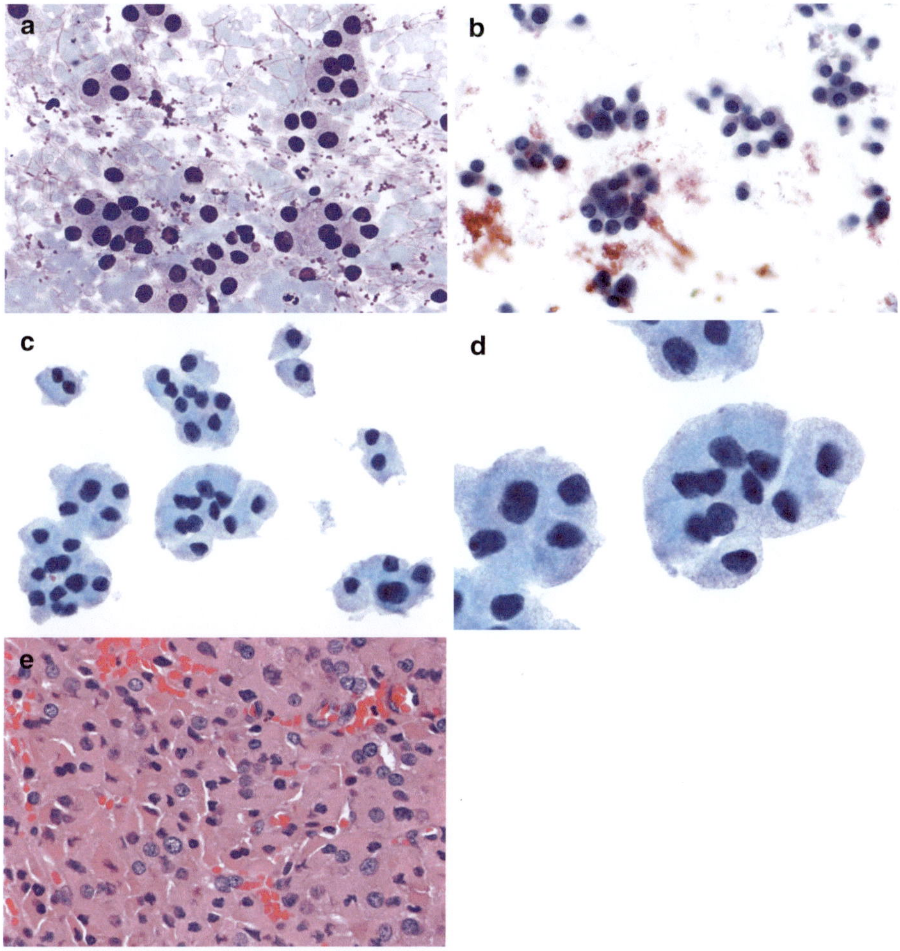

Fig. 7.9 Hurthle cell neoplasm**. (a)** Hurthle cell in a microfollicular architecture. The cytoplasm is moderately abundant, eosinophilic, and granular. Nuclei are mostly eccentric, round, and regular. Other nuclear features cannot be discerned in this preparation (DQ-stained DS). (**b**) Cytoplasm on Pap stain appears amphophilic and granular. Nuclei are small with small nucleoli (Pap-stained DS). (**c, d**) Corresponding LBP appears similar to (**b**) but with a much cleaner background and denser cytoplasm. Nucleus-to-cytoplasmic ratio is low (TP); **(e)** subsequent lobectomy showed a Hurthle cell adenoma with abundant pink granular cytoplasm (H&E). Cytological differential diagnosis of Hurthle cell neoplasm include Hurthle cell nodule in nodular hyperplasia and follicular adenomas with predominantly Hurthle cell change. If capsular or vascular invasion is histologically present, then it represents a Hurthle cell carcinoma

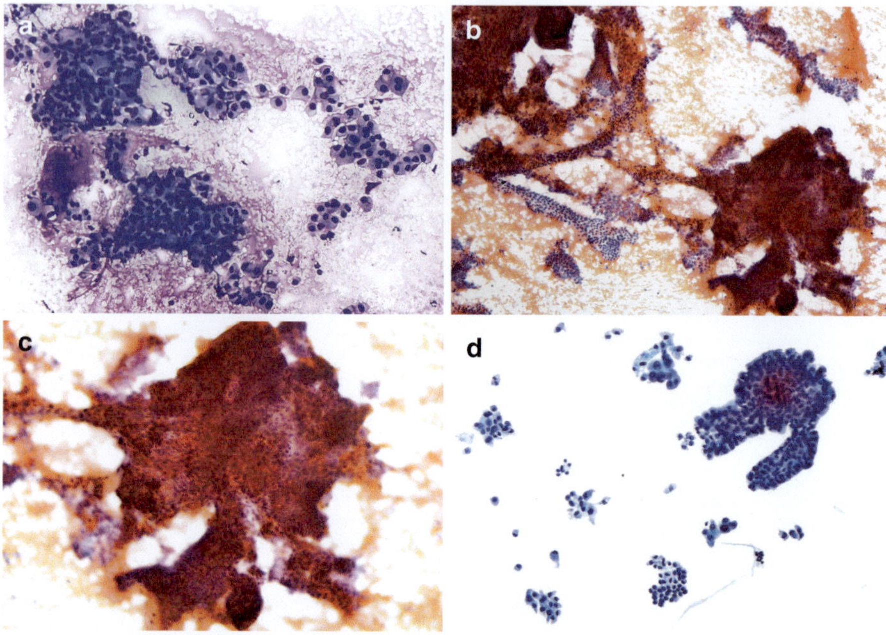

Fig. 7.10 Papillary thyroid carcinoma (PTC). (**a**) Smear is cellular, with cells in syncytia and monolayer sheets. Dense colloid is noted at the 7 o'clock position (DQ-stained DS). (**b**) Cellular smear with papillary fragments and flat syncytial sheets that are suspicious for PTC. Note the deep orangish and eosinophilic staining of tumor cells indicative of thickness of the fragment and presence of obscuring blood (Pap-stained DS). (**c**) Higher magnification of the papillary fragment shows densely packed nuclei, another cytological feature not usually seen in benign lesions (**b**, Pap-stained DS). (**d**) TP from the same case showing similar architectural arrangements as the smears, including, syncytia, flat sheets, single cells, and papillary-like groups. In TP, a central fibrovascular core usually stains as an eosinophilic centrally placed band. Nuclear features of PTC on LBP are described in images below. In this case a cytological diagnosis of PTC was rendered, and the subsequent total thyroidectomy showed classical PTC. Cases that are cytologically diagnosed as "suspicious for malignancy" in most cases show some, but not all, nuclear features of PTC

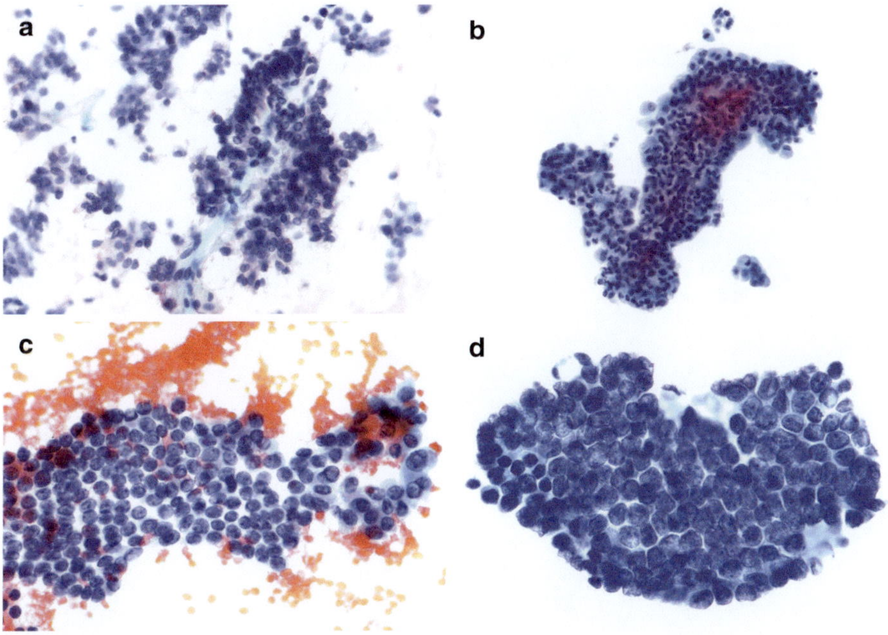

Fig. 7.11 Architectural features of PTC. (**a, b**) Papillary fragments of PTC with a fibrovascular core. Note the central eosinophilic stalk in TP (**a**, Pap-stained DS; **b**, TP). (**c**, **d**) Flat monolayered syncytial sheet of PTC in which the nuclei are prominent but the cytoplasm is indistinct. Nuclei are overlapping, round to oval with powdery chromatin, grooves, and nucleoli. Pseudoinclusions are evident in the direct smear but not in the TP (**c**, Pap-stained DS; **d**, TP)

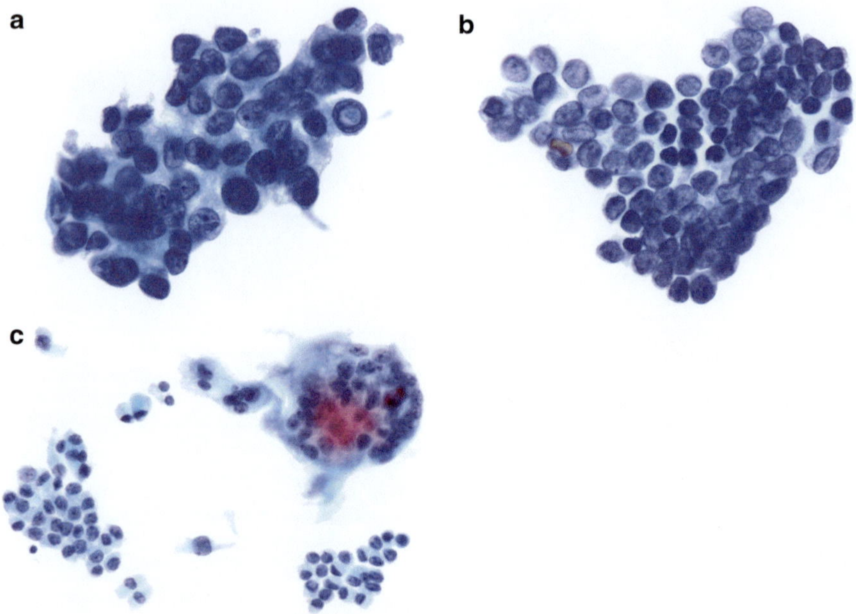

Fig. 7.12 Nuclear features of PTC on ThinPrep. (**a**, **b**), Clear nuclear morphology is seen including overlapping, round to oval shape, with powdery chromatin, grooves, and nucleoli. Pseudoinclusions are evident in image seen in (**a**). In PTC, intranuclear pseudoinclusions represent complex nuclear membrane infoldings in which the cytoplasm has projected into a portion of the nucleus; in cytology, these projections appear to be inclusions within the nucleus. Thus, the color and quality of the pseudoinclusion should match that of the cell's cytoplasm. Note this feature in image **A**. Nuclear grooves are typically seen as thin lines running longitudinally through the center of the nucleus; because the chromatin is powdery, grooves are visible as thin, dark lines matching the quality and color of the nuclear rim (**a** and **b**, TP) [11]. (**c**) Tumor cells in syncytial fragments show dense, squamoid cytoplasm. A multinucleated giant cell is seen at the right upper corner, a finding seen in both PTC and Hashimoto thyroiditis (TP)

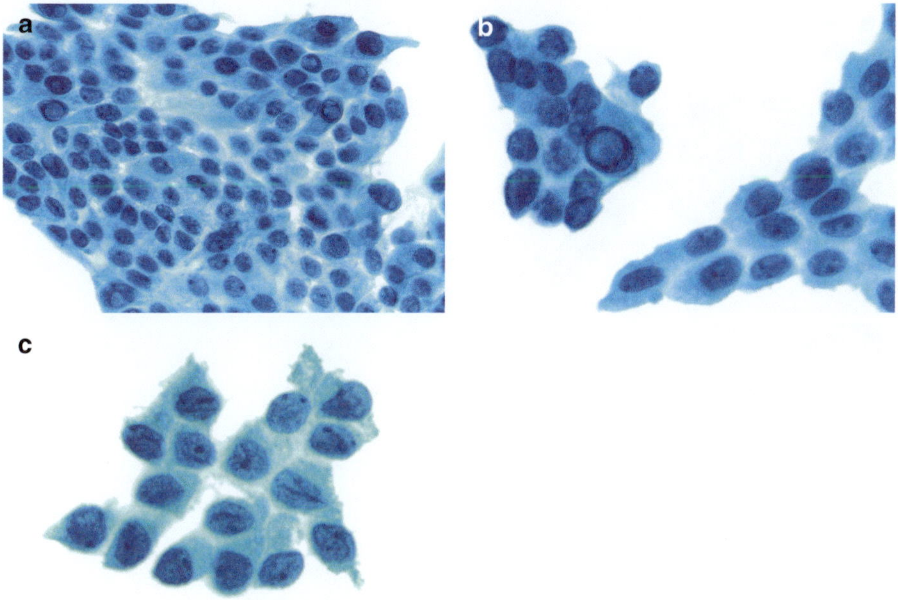

Fig. 7.13 Nuclear features of PTC on SurePath. (**a**–**c**) The cytomorphology of PTC on SP preparation is similar to that seen in TP. Note the squamoid cytoplasm and well-preserved nuclear features of PTC including round to oval shape, powdery chromatin, nucleoli, prominent pseudoinclusion, and grooves (**a**–**c**, SP). Jung et al. [12] found that for the diagnosis of PTC, classic features such as papillary fronds, nuclear pseudoinclusions, and nuclear grooves were well maintained on SP slides, and nuclear irregularity was present in every PTC case examined

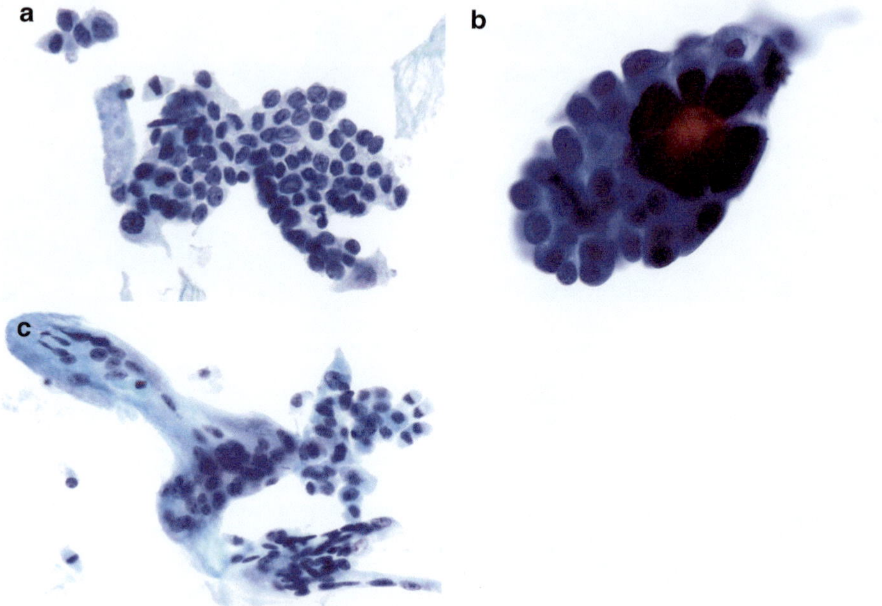

Fig. 7.14 Additional findings in PTC. (**a**) PTC can have colloid which corresponds to the "bubble gum" colloid seen in histological sections. In PTC colloid can also appear "ropy" (TP) [13]. (**b**) A psammoma body surrounded by follicular cells in a case of PTC. The follicular cells possess all nuclear features of PTC except pseudoinclusions, which were present in other cells (TP). (**c**) Irregularly shaped multinucleated giant cell in association with PTC cells (TP). The presence of multinucleated giant cell can be seen in both neoplastic and nonneoplastic conditions. Identification of nuclear features of PTC is essential for diagnosis

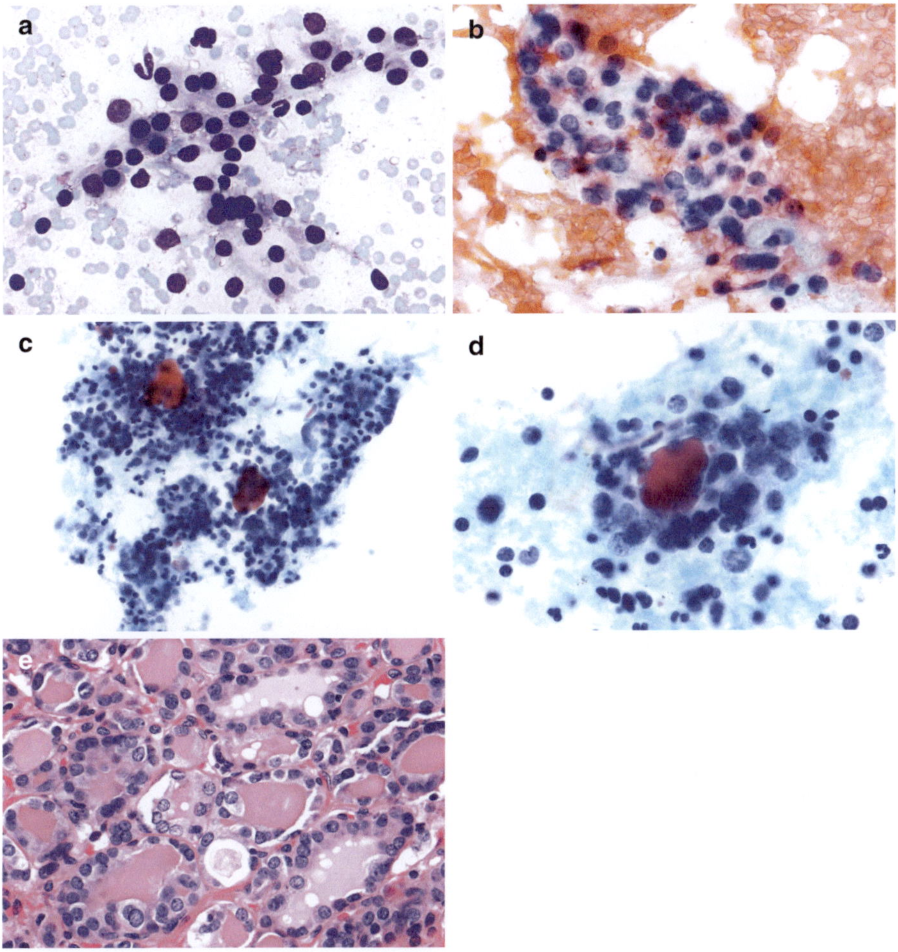

Fig. 7.15 Suspicious for malignancy. (**a**) Follicular cells are seen in microfollicles and have enlarged nuclei with some overlap (DQ-stained DS). (**b**) Follicular cells in a microfollicular architecture with enlarged, overlapping, and ovoid nuclei with powdery chromatin. However, these features are not enough for a diagnosis of PTC (Pap-stained DS). (**c**, **d**) Cellular TP preparation with numerous crowded follicular cells in a microfollicular configuration. Dense colloid is present as red-staining material contained within some follicles. The chromatin is powdery and the nuclei are enlarged and overlapping; intranuclear pseudoinclusions and grooves are not seen and a diagnosis of "suspicious for malignancy was rendered" (**c** and **d**, TP). (**e**) Subsequent lobectomy showed a follicular variant of PTC (FVPTC). Small follicles comprised of follicular cells with enlarged, overlapping nuclei and powdery chromatin. Capsular and vascular invasion was identified (H&E). The FVPTC is defined by the nuclear features of PTC in a predominantly microfollicular architecture. Invasion must be present, which cannot be assessed on cytology alone. A similar benign, noninvasive lesion is classified as "thyroid neoplasm with papillary-like nuclear features" (NIFTP). Both lesions often have a pattern of microfollicles with features of PTC on cytology and are often classified either as "indeterminate" or "suspicious for malignancy".

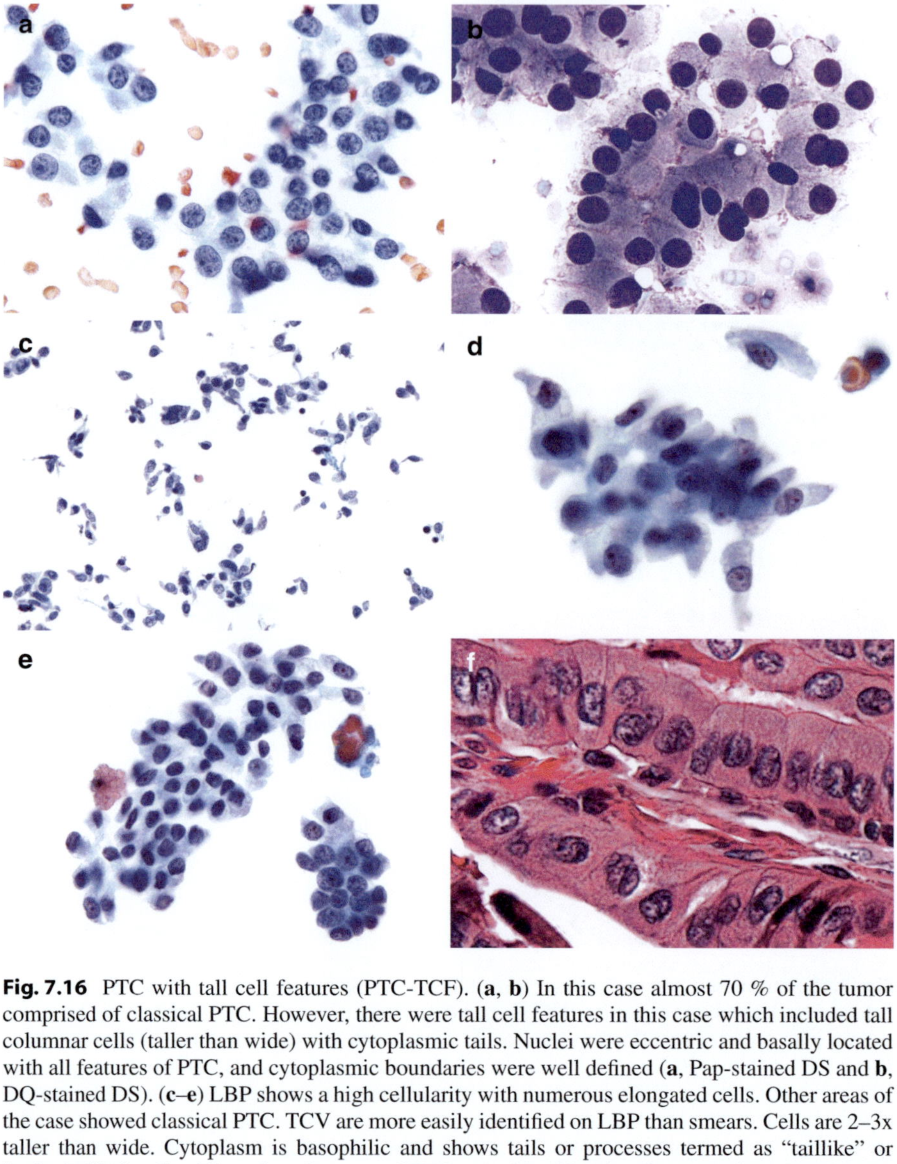

Fig. 7.16 PTC with tall cell features (PTC-TCF). (**a**, **b**) In this case almost 70 % of the tumor comprised of classical PTC. However, there were tall cell features in this case which included tall columnar cells (taller than wide) with cytoplasmic tails. Nuclei were eccentric and basally located with all features of PTC, and cytoplasmic boundaries were well defined (**a**, Pap-stained DS and **b**, DQ-stained DS). (**c–e**) LBP shows a high cellularity with numerous elongated cells. Other areas of the case showed classical PTC. TCV are more easily identified on LBP than smears. Cells are 2–3x taller than wide. Cytoplasm is basophilic and shows tails or processes termed as "taillike" or "tadpole-like" cells. All nuclear features of PTC were identified in all preparation types but are more evident in image **(e)** (**c–e**, TP); **(f)** on resection, tall cell component comprised 30 % of the tumor. The cells have a columnar shape with abundant granular cytoplasm and atypical nuclei (H&E). TCV is an aggressive variant of PTC which may demonstrate tall columnar cells with cytoplasmic tails in about one-half of FNA specimens. The neoplastic cells may also have abundant, granular cytoplasm and many intranuclear pseudoinclusions giving the nucleus a "soap-bubble" appearance

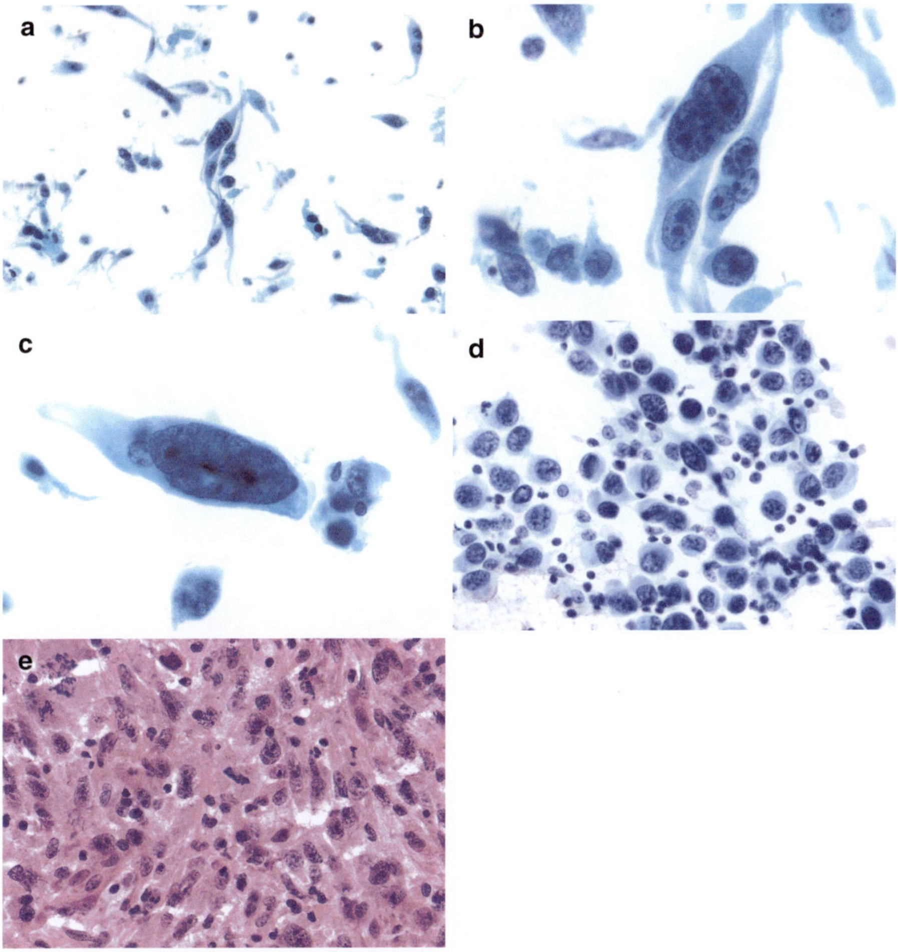

Fig. 7.17 Anaplastic thyroid carcinoma. (**a–c**) The cells have bizarre-shaped, giant nuclei with coarse chromatin and unusual cytoplasmic extensions. The clinical history of a rapidly enlarging thyroid tumor in an older person with hoarseness, dysphagia, and other compressive symptoms aids with the diagnosis (**a–c**, SP). (**d**) Same case on a smear shows pleomorphic tumor cells which appear more epithelioid while retaining the nuclear features as seen in the SP specimen (Pap-stained DS). Cellular elongation may be more apparent in SP due to technical reasons (see Chap. 1). Anaplastic thyroid carcinoma is an uncommon and extremely aggressive variant of thyroid carcinoma; cells often resemble a high-grade carcinoma and do not have features distinctive of a thyroid origin. Cells demonstrate great pleomorphism and may also have squamous features in some instances. Thyroid-specific markers such as thyroglobulin and TTF-1 are often not expressed. (**e**) Subsequent resection of anaplastic thyroid carcinoma with a sea of malignant and anaplastic tumor cells, with numerous mitotic figures (H&E).

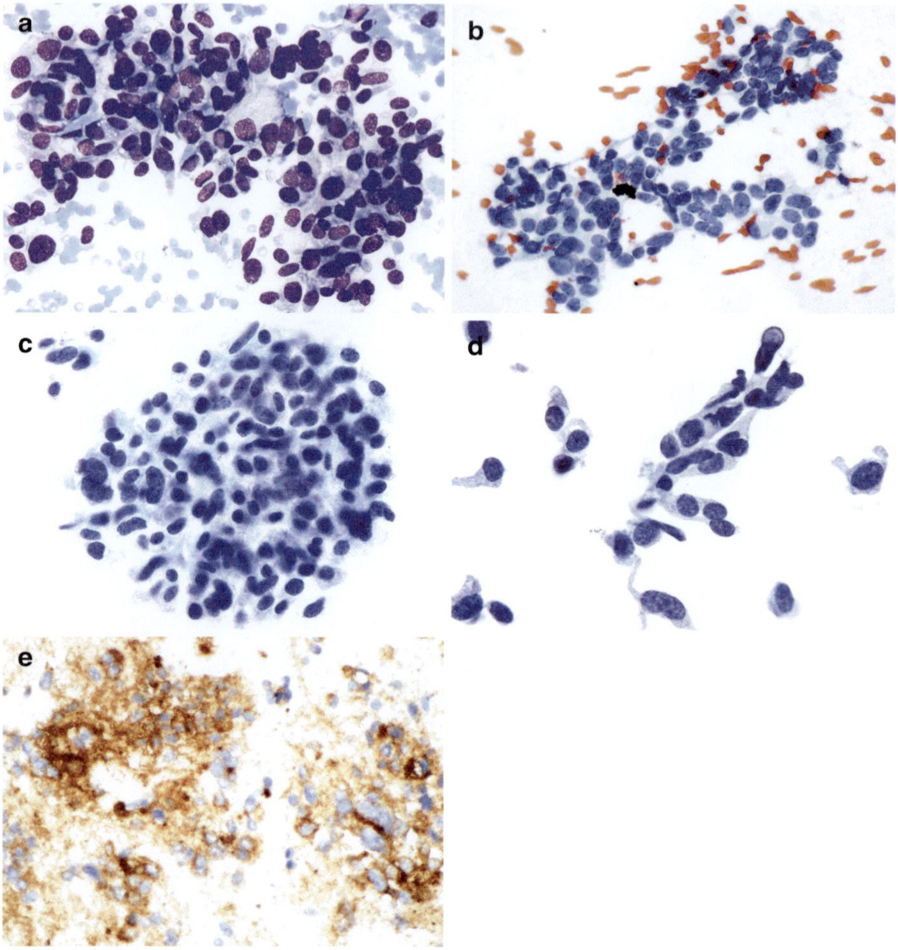

Fig. 7.18 Medullary thyroid carcinoma. (**a–d**) All preparations, including DQ- and Pap-stained smears and TP, show similar cytological features. Tumor cells demonstrate neuroendocrine features including discohesive, plasmacytoid cells with a "salt and pepper" chromatin pattern. Nuclei are mostly oval to spindled with regular borders and demonstrate anisonucleosis. Cytoplasm is scant, basophilic, and granular (**a**, DQ-stained DS; **b**, Pap-stained DS; **c** and **d**, TP). (**e**), Positive calcitonin stain along with the cytological features and an elevated serum calcitonin level confirmed the diagnosis of medullary carcinoma. Tumor cells also expressed neuroendocrine markers including synaptophysin and chromogranin (calcitonin stain). Resection showed medullary carcinoma.

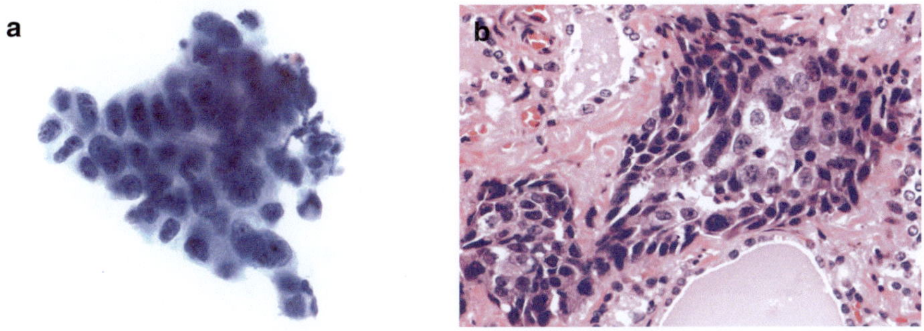

Fig. 7.19 Metastatic squamous cell carcinoma. (**a**) This patient had a history of nonkeratinizing squamous cell carcinoma of the lung. Tumor cells have dense, rigid, and well-defined cytoplasm and prominent nucleoli. A mitotic figure is also present (TP). (**b**) Resection specimen shows non-keratinizing squamous cell carcinoma, infiltrating between normal benign thyroid follicles (H&E). The most frequent metastases to the thyroid gland are primary tumors from the kidney, lung, and head and neck region [14]. Nonkeratinizing squamous cell carcinoma looks similar to that found at other sites of the body, with centrally located, irregular-shaped nuclei and dense cytoplasm. The differential diagnosis includes anaplastic thyroid carcinoma. Patients with metastatic squamous cell carcinoma typically have a known primary, making the diagnosis less challenging.

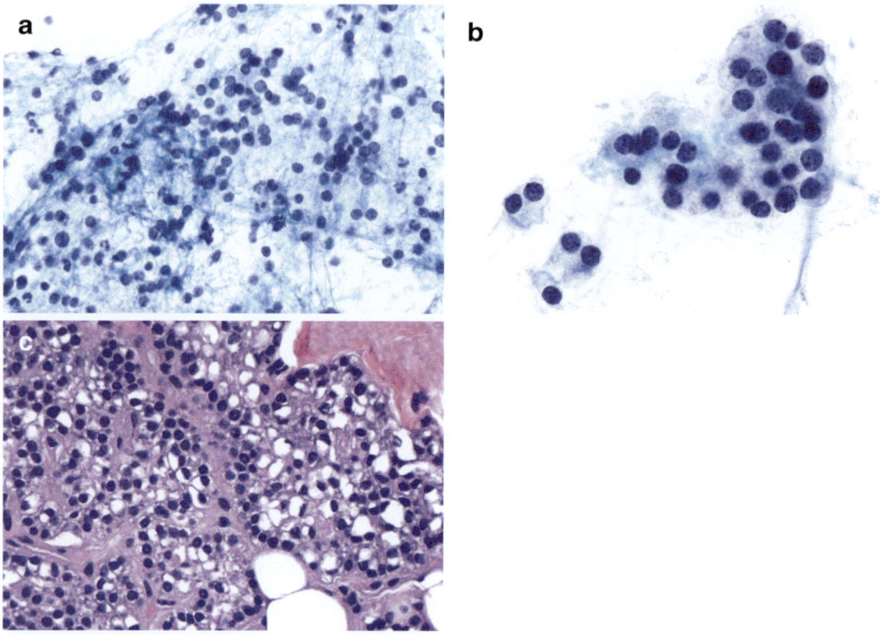

Fig. 7.20 Parathyroid hyperplasia. (**a–b**) Parathyroid cells have the same cytomorphology as neuroendocrine cells in other locations; the cells may be singly dispersed and/or in fragments with microfollicular architecture. Nuclei are eccentrically placed; small, round, and monotonous cytoplasm is abundant and granular. On Pap stain, the chromatin has the typical "salt and pepper" appearance. Parathyroid cells may resemble thyroid follicular cells and thus may be diagnosed as an "atypical follicular lesion" if their parathyroid origin is not suspected. Immunostaining for parathyroid hormone (PTH) can confirm their origin; patients should have elevated PTH levels. If on-site adequacy is performed, an additional needle pass can be taken and sent to the laboratory to measure PTH levels (**a**, Pap-stained DS; **b**, TP). (**c**) Corresponding excision of hyperplastic parathyroid glands; the cells are present in a ribbon-like appearance (H&E). Hyperplastic parathyroid gland may present as an intra- or extra-thyroidal nodule, which may be sampled.

Suggested Reading

1. Cibas ES, Ali SZ. NCI thyroid FNA state of the science conference. The Bethesda system for reporting thyroid cytopathology. Am J Clin Pathol. 2009;132:658–65.
2. Malle D, Valeri RM, Pazaitou-Panajiotou K, Kiziridou A, Vainas I, Destouni C. Use of a thin-layer technique in thyroid fine needle aspiration. Acta Cytol. 2006;50:23–7.
3. Michael CW, Hunter B. Interpretation of fine-needle aspirates processed by the ThinPrep technique: cytologic artifacts and diagnostic pitfalls. Diagn Cytopathol. 2000;23:6–13.
4. Tulecke MA, Wang HH. ThinPrep for cytologic evaluation of follicular thyroid lesions: correlation with histologic findings. Diagn Cytopathol. 2004;30:7–13.
5. Rossi ED, Raffaelli M, Zannoni GF, Pontecorvi A, Mulè A, Callà C, Lombardi CP, Fadda G. Diagnostic efficacy of conventional as compared to liquid-based cytology in thyroid lesions: evaluation of 10,360 fine needle aspiration cytology cases. Acta Cytol. 2009;53:659–66.
6. Nishino M. Molecular cytopathology for thyroid nodules: a review of methodology and test performance. Cancer Cytopathol. 2016;124:14–27.
7. Krane JF, Cibas ES, Alexander EK, Paschke R, Eszlinger M. Molecular analysis of residual ThinPrep material from thyroid FNAs increases diagnostic sensitivity. Cancer Cytopathol. 2015;123:356–61.
8. Kim Y, Choi KR, Chae MJ, Shin BK, Kim HK, Kim A, Kim BH. Stability of DNA, RNA, cytomorphology, and immunoantigenicity in residual ThinPrep specimens. APMIS. 2013;121:1064–72.
9. Eszlinger M, Piana S, Moll A, Bösenberg E, Bisagni A, Ciarrocchi A, Ragazzi M, Paschke R. Molecular testing of thyroid fine-needle aspirations improves presurgical diagnosis and supports the histologic identification of minimally invasive follicular thyroid carcinomas. Thyroid. 2015;25:401–9.
10. Faquin WC, Cibas ES, Renshaw AA. "Atypical" cells in fine-needle aspiration biopsy specimens of benign thyroid cysts. Cancer. 2005;105:71–9.
11. Geers C, Bourgain C. Liquid-based FNAC of the thyroid: a 4-year survey with SurePath. Cancer Cytopathol. 2011;119:58–67.
12. Jung CK, Lee A, Jung ES, Choi YJ, Jung SL, Lee KY. Split sample comparison of a liquid-based method and conventional smears in thyroid fine needle aspiration. Acta Cytol. 2008;52:313–9.
13. Duncan LD, Forrest L, Law Jr WM, Hubbard E, Stewart LE. Evaluation of thyroid fine-needle aspirations: can ThinPrep be used exclusively to appropriately triage patients having a thyroid nodule? Diagn Cytopathol. 2011;39:341–8.
14. Hegerova L, Griebeler ML, Reynolds JP, Henry MR, Gharib H. Metastasis to the thyroid gland: report of a large series from the Mayo clinic. Am J Clin Oncol. 2015;38:338–42.
15. Lee SH, Jung CK, Bae JS, Jung SL, Choi YJ, Kang CS. Liquid-based cytology improves preoperative diagnostic accuracy of the tall cell variant of papillary thyroid carcinoma. Diagn Cytopathol. 2014;42:11–7.

Fine Needle Aspiration of Salivary Gland

8

William C. Faquin and Mine Onenerk

Introduction

The salivary glands are comprised of three major (parotid, submandibular, and sublingual) and numerous minor salivary glands which are dispersed within the upper aerodigestive tract. Combined with clinical and radiologic evaluation, fine needle aspiration (FNA) represents a useful initial procedure in the assessment of a salivary gland mass. The normal salivary gland parenchyma is composed of small exocrine units consisting of acini and a ductal system (intercalated, striated, and excretory ducts in the major glands). The serous or mucinous content of the acinar cells and length of the ducts show site-specific differences.

Salivary gland tumors constitute a heterogeneous group with more than 40 primary benign and malignant neoplasms described by the WHO. Although some exceptions exist, there is a female predominance and mean overall age of 46–47 years for patients presenting with a salivary gland tumor. Benign tumors outnumber malignant, and pleomorphic adenoma (PA) is the most common type of salivary gland neoplasm in the adult population.

Except for PA, the prevalence of benign and malignant tumors is reported to be highly variable among different populations. In most of the series, Warthin tumor is the second most common benign tumor, and it is almost exclusively seen in the parotid gland. Mucoepidermoid carcinoma (MEC), polymorphous low-grade adenocarcinoma, adenoid cystic carcinoma, and acinic cell carcinoma comprise the more common malignant tumors.

Several studies have shown that the majority of primary salivary gland tumors occur in the parotid gland (64–80 %). Tumor prevalence is relatively low in other sites – approximately 10 % for submandibular gland, 1 % for sublingual gland, and 9–23 % for minor salivary glands. The rate of malignancy is inversely proportional to the gland size: 15–32 % for the parotid gland, 41–45 % for the submandibular gland, 70–90 % for the sublingual gland, and 90 % for the minor salivary glands. Metastatic tumors comprise only 5 % of all malignant salivary gland tumors, and the majority is squamous cell carcinomas. FNA can also be performed for nonneoplastic mass-forming salivary gland lesions such as acute or chronic sialoadenitis

R.S. Hoda et al., *Diagnostic Liquid-Based Cytology*,
DOI 10.1007/978-3-662-53905-7_8, © Springer-Verlag GmbH Germany 2017

and granulomatous sialoadenitis. Cytological diagnoses are based on cytomorphologic criteria. Ancillary tests including immunohistochemistry, histochemistry, flow cytometry, and molecular analysis (e.g., FISH, NGS, PCR) can contribute to improved cytologic accuracy. Surgery is the treatment option for almost all primary salivary gland tumors, but the extent of surgery, facial nerve conservation, and performance of a neck dissection will vary depending on the preoperative diagnosis. Thus, an accurate FNA diagnosis is useful for guiding the preoperative management strategy.

Cytological Reporting Guidelines

Currently, a uniform reporting system is lacking for salivary gland FNAs. Most laboratories use conventional cytologic categories to report results (nondiagnostic, negative for malignant cells, atypical, suspicious for malignancy, or positive for malignant cells) and a specific (pleomorphic adenoma, Warthin tumor, acinic cell carcinoma, etc.) or descriptive diagnosis (basaloid or myoepithelial neoplasm, etc.) in their reports. There have been efforts to introduce a more reproducible and risk-based classification system known as the Milan System for Reporting Salivary Gland Cytopathology.

The Milan System is a six-tiered system with the following proposed categories: Non-diagnostic, Non-neoplastic, Atypia of Undetermined Significance (AUS), Neoplastic, Suspicious for Malignancy and Malignant. It is important in the 'Suspicious for Malignancy and the Malignant' categories to distinguish low-grade from high-grade tumors.

Indication, Collection, and Laboratory Processing of Cytological Samples

Salivary gland FNA is an accurate and cost-effective method in the preoperative management of patients with a salivary gland mass. The overall accuracy ranges from 81 to 98 % and is higher for benign tumors when compared to malignant. An accurate assessment of salivary gland cytology depends on the adequacy/preparation quality of the specimen and the cytopathologists' awareness of overlapping features of different tumor types and diverse cytomorphologic appearances of any given specific tumor.

Methods of Specimen Collection

An FNA of a salivary gland mass can be performed by using a 23–27-gauge needle with or without applying negative pressure. Air-dried and alcohol-fixed smears as well as liquid-based slides can give complementary information in the characterization of cytologic features. For example, air-dried smears are especially valuable for appreciating the different stromal and matrix characteristics of various tumor types which can be a significant diagnostic clue to the correct diagnosis. Correlation of the cytologic specimen and radiologic findings is essential for an accurate diagnosis and can be integrated when the FNA is performed under ultrasound guidance.

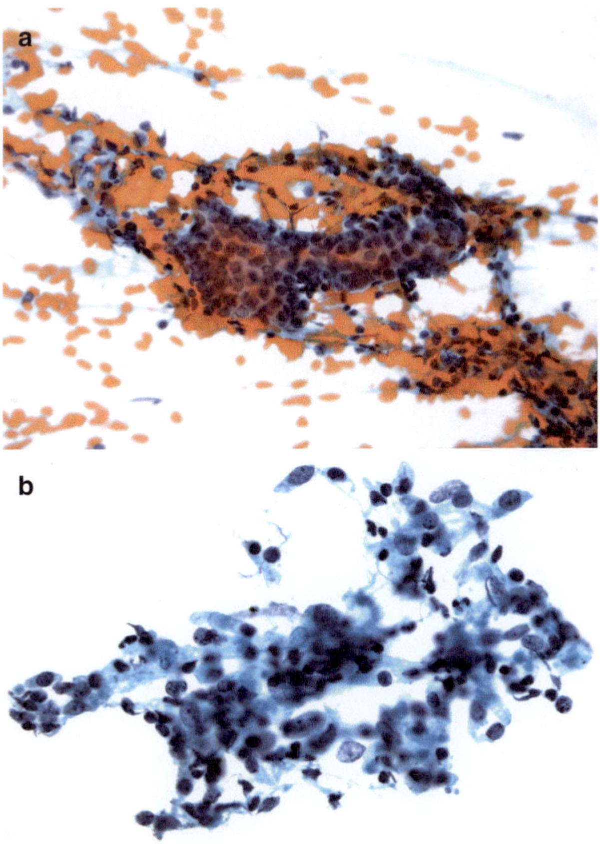

Fig. 8.1 Chronic sialadenitis of the submandibular gland. (**a**) Aspirations from chronic sialadenitis are generally hypocellular. In this case ductal cells form a flat sheet which resembles a tubule. Cells have evenly spaced nuclei, an important finding supporting the benign nature of the ductal cells. Ductal nuclei are slightly enlarged and uniform with fine chromatin. Acinar cells are generally rarely encountered in chronic sialadenitis. Inflammatory cells are present in the background, and they rarely infiltrate ductal cells (Pap-stained DS). (**b**) Inflammation is more easily recognized in this slide. Histiocytes and lymphocytes are admixed with dispersed ductal cells (SP). (**c**) Lymphocytes infiltrate around the intercalated ducts. Fibrosis around the ducts and acinar atrophy are also present (H&E)

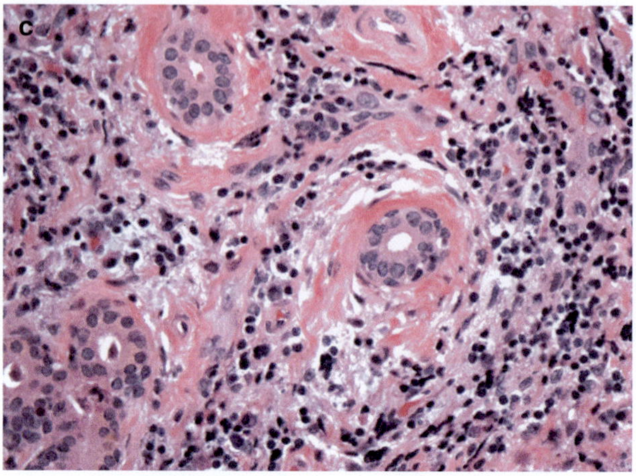

Fig. 8.1 (continued)

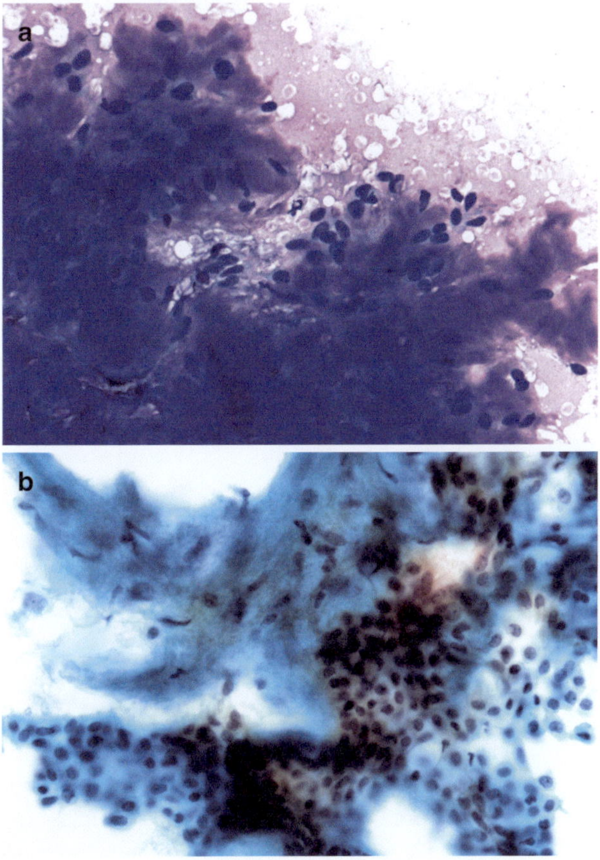

Fig. 8.2 Pleomorphic adenoma (PA) of the parotid gland. (**a**) Epithelial and myoepithelial cells are admixed with characteristic metachromatic fibrillary matrix with frayed edges (Giemsa-stained DS). Romanovsky-type stains are more valuable for the evaluation of the matrix characteristics. The fibrillary consistency of the matrix is important in the differentiation of PA from other salivary gland neoplasms such as adenoid cystic carcinoma or basal cell adenoma. (**b**) Myoepithelial cells are embedded within the fibrillary matrix. Neoplastic cells have moderate amounts of cytoplasm, nuclei are uniform and bland, and the chromatin is fine and evenly dispersed (SP). (**c, d**) A different PA case with fibrillary matrix and cells with round to oval nuclei and moderate amounts of cytoplasm (**c**, Pap-stained DS; **d**, TP). (**e**) The biphasic growth pattern with ducts lined by cuboidal epithelial cells and intervening cellular myxoid stromal component (H&E)

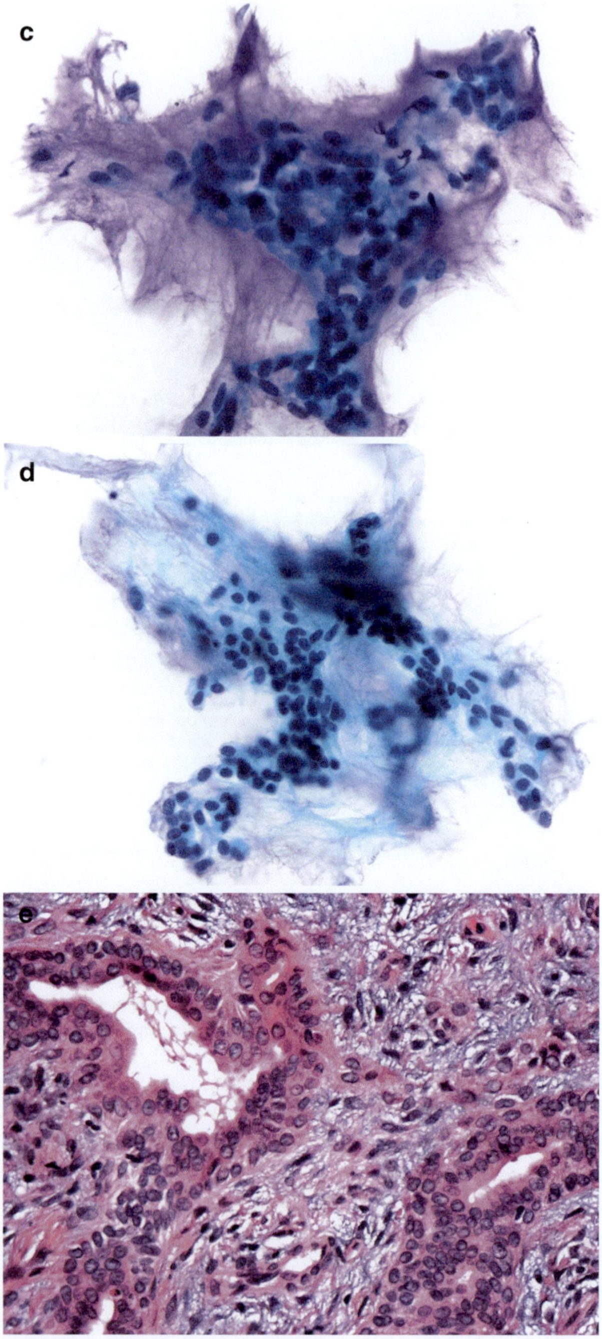

Fig. 8.2 (continued)

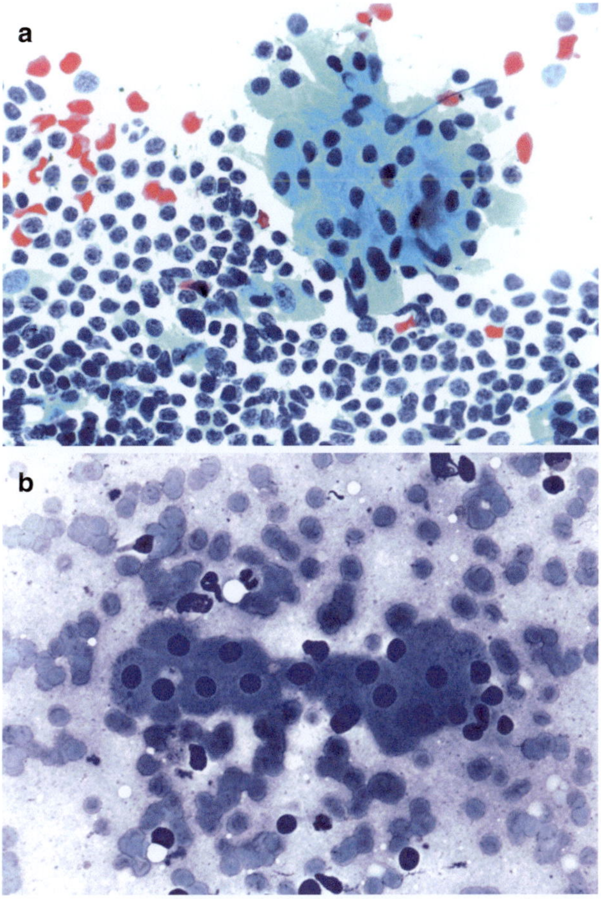

Fig. 8.3 Warthin tumor of the parotid gland. (**a**) A flat sheet of oncocytic cells and a lymphocytic background are characteristic. Oncocytes have abundant cytoplasm and distinct cell borders. Nuclei are round and centrally located with prominent nucleoli (Pap-stained DS). (**b**) Oncocytic cells arranged in a cohesive sheet without crowding. Absence of cellular crowding is an important feature when differentiating Warthin tumor from other types of tumors which also have oncocytes such as oncocytic variant of mucoepidermoid carcinoma. Oncocytes have "waxy" cytoplasm which can be better appreciated in Romanovsky-type stains. Dense rather than a granular consistency of the cytoplasm is important to discriminate these cells from acinar cells. Scant lymphocytes are seen in the clean background (Giemsa-stained DS). (**c**) Uniform oncocytic cells are forming cohesive sheets. Cytoplasmic borders are distinct and some cells have prominent nucleoli (SP). (**d**) Bilayered epithelium with columnar cells located in the luminal parts and less obvious basal cells surrounded by lymphocytic stroma. Both luminal and basal cells have finely granular cytoplasm, distinct cell membranes, uniform nuclei, and prominent nucleoli (H&E)

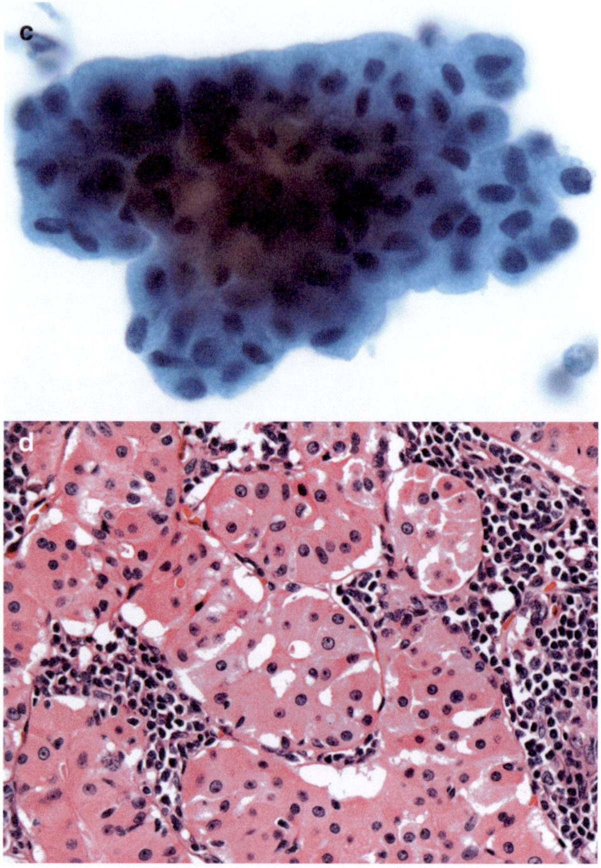

Fig. 8.3 (continued)

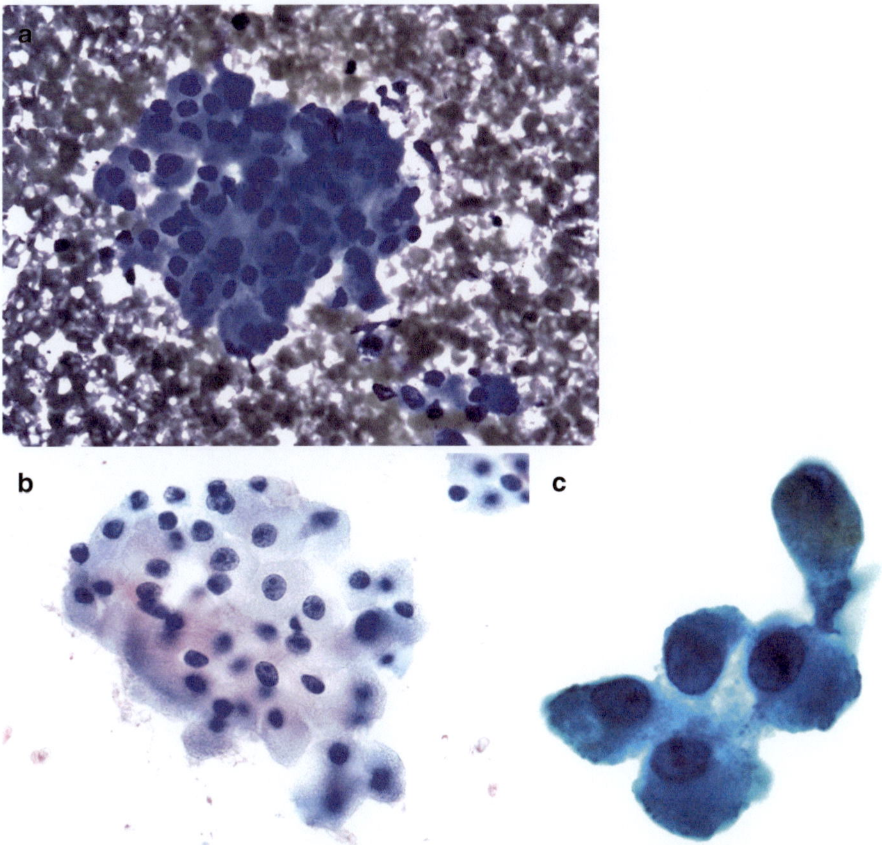

Fig. 8.4 Oncocytoma of the parotid gland. (**a**) Three-dimensional clusters of oncocytic cells. Cells have abundant "waxy" cytoplasm. The background is clean without lymphocytes. Oncocytes can be seen in a broad spectrum of benign and malignant lesions (Warthin tumor, mucoepidermoid carcinoma, oncocytic carcinoma) as well as in oncocytic hyperplasia and metaplasia which cannot be discriminated on cytology (Giemsa-stained DS). (**b**) Oncocytes with distinct cell borders and bland nuclei with prominent nucleoli. Texture of the oncocytic cytoplasm is different in Romanovsky-type and Papanicolaou-stained slides. It is nongranular and "waxy" in the former and "densely granular" in the latter (Pap-stained DS). (**c**) Single oncocytic cells in a clear background (SP). (**d**) Oncocytic cells have intensely eosinophilic cytoplasm with round-oval nuclei and prominent nucleoli. Cells are monotonous without atypia (H&E)

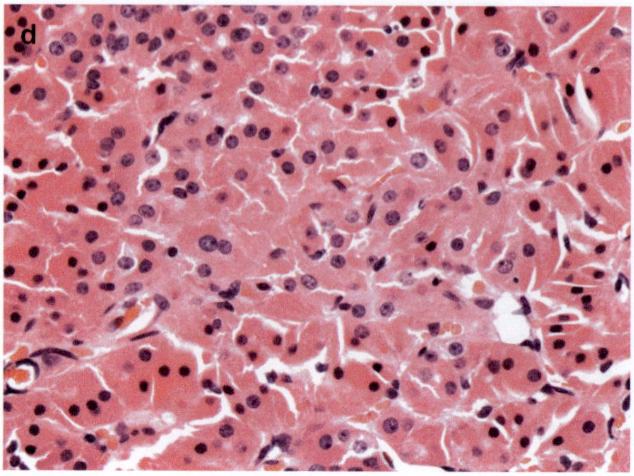

Fig. 8.4 (continued)

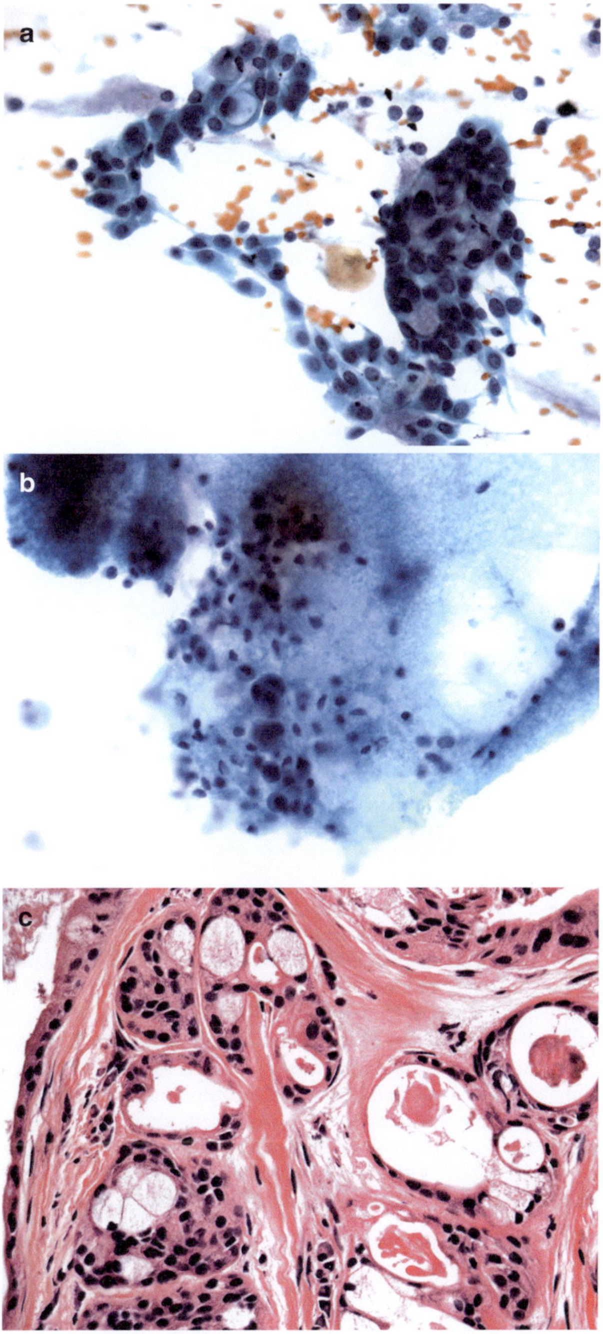

Fig. 8.5 Low-grade mucoepidermoid carcinoma of the parotid gland. (**a**) Loose clusters of admixed intermediate cells, mucus, and epidermoid cells. Mucus cells have a signet ring-like appearance. Epidermoid cells are squamoid with cyanophilic cytoplasm. Intermediate cells are smaller and constitute the major cellular component in the slide. Scant mucoid material can be seen at the periphery of the cell groups (Pap-stained DS). (**b**) Intermediate cells within mucoid material. Although finding a mucocyte is highly suggestive of mucoepidermoid carcinoma, the intermediate cells are sometimes the predominant component in aspirations of low to intermediate grade tumors. Cells are generally bland to mildly atypical but are not overtly malignant (SP). (**c**) Low-grade mucoepidermoid carcinoma with cystic spaces. Cyst-lining intermediate cells with dark staining round nuclei and moderate amounts of eosinophilic cytoplasm. Mucin-containing cells (mucocytes) with pale abundant cytoplasm can be observed. Cells have uniform nuclei without significant atypia (H&E)

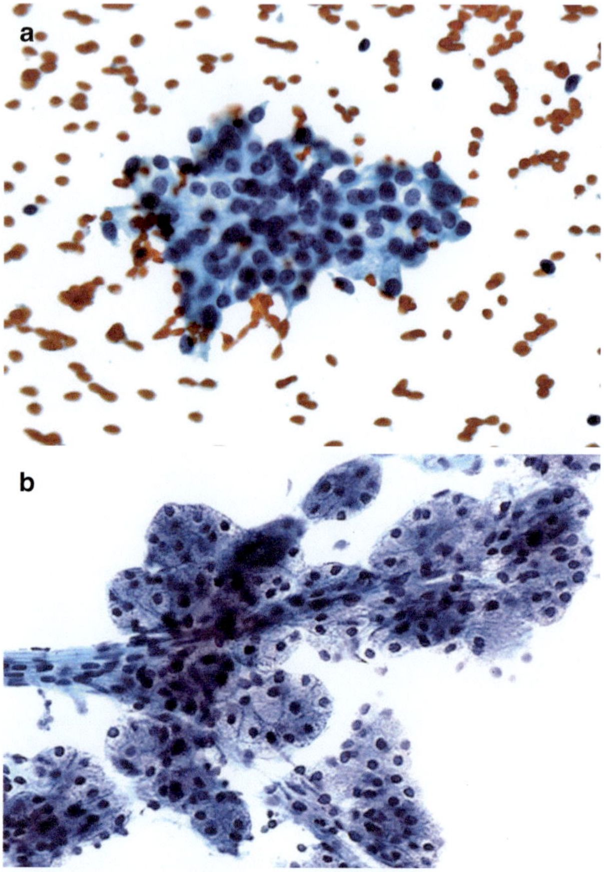

Fig. 8.6 Acinic cell carcinoma of the parotid gland. (**a**) Tumor cells form loose cohesive groups unlike the cohesive grape-like clusters of normal salivary gland acinar cells. The tumor consists of fairly uniform polygonal cells with abundant, finely vacuolated cytoplasm which has a slightly granular and basophilic quality. Absence of organized ductal cells and adipose tissue is an important clue when differentiating acinic cell carcinoma from normal tissue. Nuclei are slightly enlarged and round but bland without significant atypia. The background is clean with rare single cells and occasional stripped nuclei (Pap-stained DS). (**b**) Normal salivary gland aspiration for comparison (Pap-stained DS). (**c**) The cytoplasm of the tumor cells is abundant and has a somewhat flocculent appearance. The serous cells show variably enlarged nuclei and mild pleomorphism. Chromatin is open and prominent nucleoli can be observed in some cells (SP). (**d**) Tumor composed of serous acinar cells with abundant coarse cytoplasmic zymogen granules. Thin fibrous septa can be observed between the cell groups (H&E)

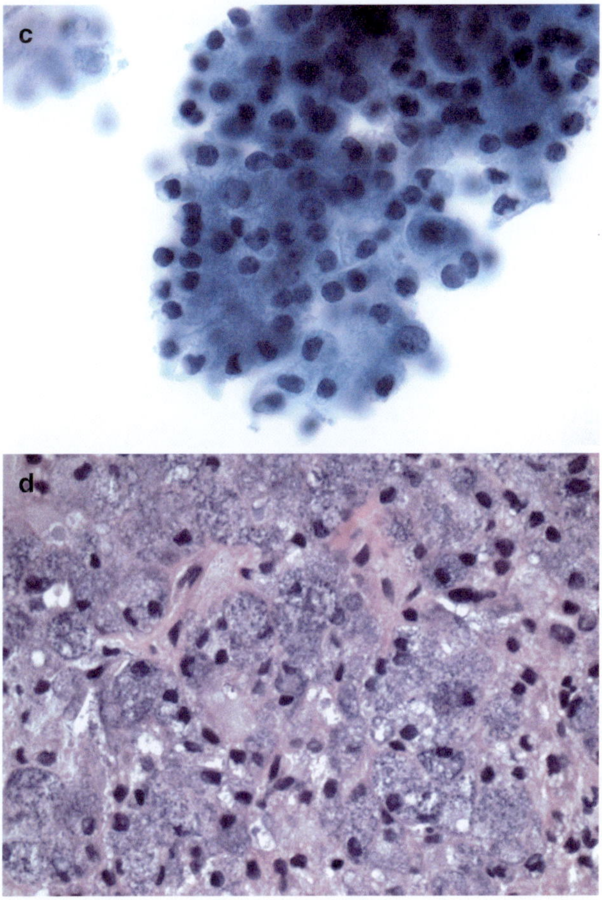

Fig. 8.6 (continued)

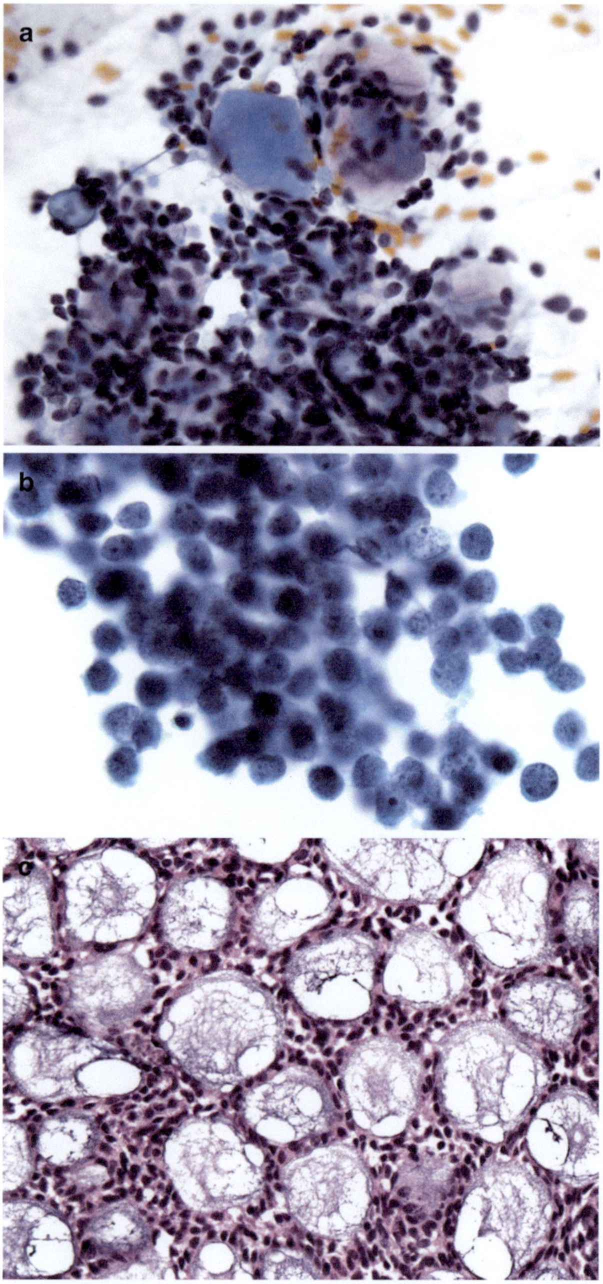

Fig. 8.7 Adenoid cystic carcinoma of the parotid gland. (**a**) Small basaloid tumor cells with hyperchromatic nuclei and scant cytoplasm surrounding hyaline globular matrix. The matrix is acellular and has sharp borders in contrast with the fibrillary matrix of pleomorphic adenoma with frayed edges (Pap-stained DS). (**b**) Tumor cells with scant cytoplasm and small nucleoli. Adenoid cystic carcinoma has three architectural growth patterns – cribriform, tubular, and solid. The solid type usually does not have a prominent matrix component which makes its distinction on cytology difficult from other basaloid tumors such as basal cell adenoma or cellular pleomorphic adenoma. However, this differential diagnosis has clinical significance because adenoid cystic carcinoma is an aggressive tumor with a tendency to invade nerves (SP). (**c**) Adenoid cystic carcinoma, cribriform type. A sheetlike arrangement of cribriform spaces with delicate violet-staining matrix and surrounded by small uniform basaloid cells (H&E)

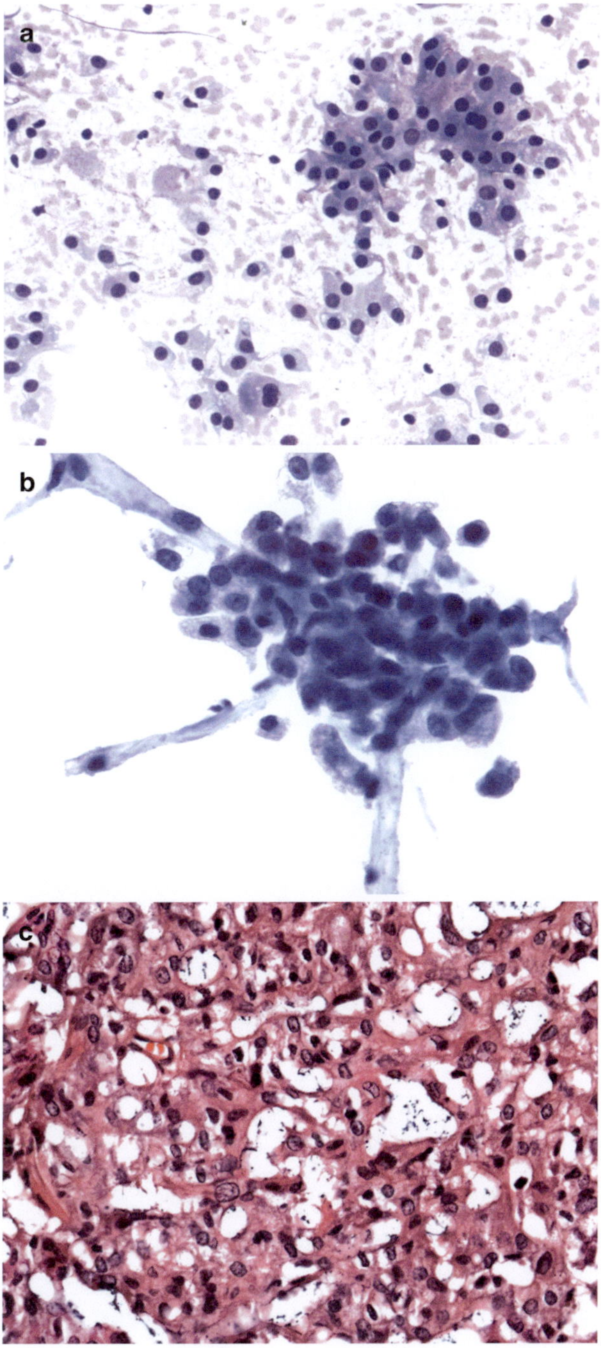

Fig. 8.8 Mammary analogue secretory carcinoma of salivary gland origin (MASC) of the parotid gland. (**a**) Tumor cells form loose clusters or are dispersed as single cells. Cells are uniform with large amounts of vacuolated cytoplasm and round eccentric nuclei. These cells can resemble the neoplastic cells of acinic cell carcinoma. However, in MASC, the cells do not contain cytoplasmic zymogen granules, and mucoid secretory material can be seen in the background (Giemsa-stained DS). (**b**) Crowded cluster of polygonal cells with abundant, vacuolated cytoplasm, indistinct cell borders, uniform nuclei, and distinct nucleoli (TP). (**c**) MASC showing a microcystic and solid architecture. Cells have abundant eosinophilic cytoplasm, vesicular nuclei, and distinct nucleoli (H&E)

Suggested Reading

1. Parfitt JR, McLachlin CM, Weir MM. Comparison of ThinPrep and conventional smears in salivary gland fine-needle aspiration biopsies. Cancer. 2007;111:123–9.
2. Al-Khafaji BM, Afify AM. Salivary gland fine needle aspiration using the ThinPrep technique: diagnostic accuracy, cytologic artifacts and pitfalls. Acta Cytol. 2001;45:567–74.
3. Rarick JM, Wasman J, Michael CW. The utility of liquid-based cytology in salivary gland fine-needle aspirates: experience of an academic institution. Acta Cytol. 2014;58:552–62.
4. Stewart CJ, MacKenzie K, McGarry GW, Mowat A. Fine-needle aspiration cytology of salivary gland: a review of 341 cases. Diagn Cytopathol. 2000;22:139–46.
5. Hughes JH, Volk EE, Wilbur DC. Pitfalls in salivary gland fine-needle aspiration cytology: lessons from the College of American pathologists interlaboratory comparison program in nongynecologic cytology. Arch Pathol Lab Med. 2005;129:26–31.
6. Hipp J, Lee B, Spector ME, Jing X. Diagnostic yield of ThinPrep preparation in the assessment of fine-needle aspiration biopsy of salivary gland neoplasms. Diagn Cytopathol. 2015;43:98–104.
7. Anderson CE, Duvall E, Wallace WA. A single ThinPrep® slide may not be representative in all head and neck fine needle aspirate specimens. Cytopathol. 2009;20:87–90.
8. Layfield LJ, Glasgow BJ. Diagnosis of salivary gland tumors by fine-needle aspiration cytology: a review of clinical utility and pitfalls. Diagn Cytopathol. 1991;7:267–72.

Fine Needle Aspiration of the Lung

9

Introduction

Lung cancer is the most common malignancy in the United States, and malignancies from other sites of the body frequently metastasize to the lung. In addition, the increased utilization of radiologic studies over time has led to a corresponding increase in the discovery of incidental lung lesions ("incidentalomas"). The advent of endoscopic ultrasound-guided bronchoscopy and, more recently, navigational bronchoscopy has allowed sampling of these lesions by fine needle aspiration (FNA). In addition, mediastinal lymph nodes may be sampled to determine the presence of metastases and allow staging of the patient by FNA without the need for mediastinoscopy, a more invasive procedure. In patients who are poor surgical candidates, FNA can be used to obtain malignant cells for molecular studies and allowing eligible patients to receive targeted therapies which may extend their lifespan.

As described in Chap. 6, multiple modalities (e.g., bronchial brushings, bronchoalveolar lavage, etc.) may be used to investigate lung lesions. Fine needle aspiration complements these methods and can also obtain diagnostic material when tissue biopsy is not possible or inadequate. In addition to the use of bronchoscopic procedures, transcutaneous FNA may be employed to sample peripheral or pleural-based lesions. Awareness of the procedure used to obtain material is critical, as it often determines the type of background contamination encountered; for instance, a bronchoscopic sampling may be contaminated by bronchial respiratory epithelial cells.

Because bronchoscopy is a time-consuming and expensive procedure, rapid on-site evaluation (ROSE) for adequacy is often performed to reduce the probability the procedure may need to be repeated due to an insufficient or otherwise nondiagnostic sample. In such instances, direct smears (DS) are prepared, stained with Diff-Quik (DQ), and evaluated on-site. However, needle rinses or dedicated needle passes may be placed in liquid-based preservative to allow for additional morphologic evaluation or molecular testing once the sample has been returned to the

R.S. Hoda et al., *Diagnostic Liquid-Based Cytology*,
DOI 10.1007/978-3-662-53905-7_9, © Springer-Verlag GmbH Germany 2017

laboratory. This can ensure well-preserved cells to be evaluated on the Pap stain and compared to those seen on-site. Due to the technical nature of bronchoscopy, it may be difficult for specimens to be smeared rapidly, which can result in premature clot formation and air-drying artifact, both of which limit morphologic interpretation. Because material can be easily collected in liquid-based preparation (LBP) preservative, this method can reduce such artifacts.

At sites lacking a pathologist or cytotechnologist for ROSE, the bronchoscopist may collect needle passes in LBP medium. This allows the bronchoscopist to focus on the procedure and patient, reduces the amount of time required to smear slides, and reduces the possibility of clot formation and/or air-drying artifact. A single liquid-based slide can be prepared from multiple passes, reducing the amount of time required to screen and diagnose each specimen. It also eliminates the problem of glass slides being transported to the laboratory, which may break in transit.

Studies have shown that TP can have improved accuracy and efficiency in diagnosing malignant lung lesions on FNA as compared to the use of conventional smears. For example, Konofaos et al. [1] found a diagnosis could be established in 75/80 patients, as compared to 54/80 patients in which conventional smears were used; the conventional smear technique also had a significantly greater rate of inadequate material.

A number of clinically relevant molecular alterations in lung cancer – in particular, lung adenocarcinoma – have been described. The most common of these include mutations in *KRAS*, *EGFR*, and *BRAF* and translocations in *ALK*. Standard of care requires that any potential lung adenocarcinoma be tested for these alterations to determine individual-based therapy. Studies have shown that material stored in LBP preservative is adequate for such testing [2–4]. Because the glass slide preparation does not consume the entire sample, residual sample can be used for molecular testing. In these instances, the molecular test results will correlate directly with the material seen on the glass slide preparations.

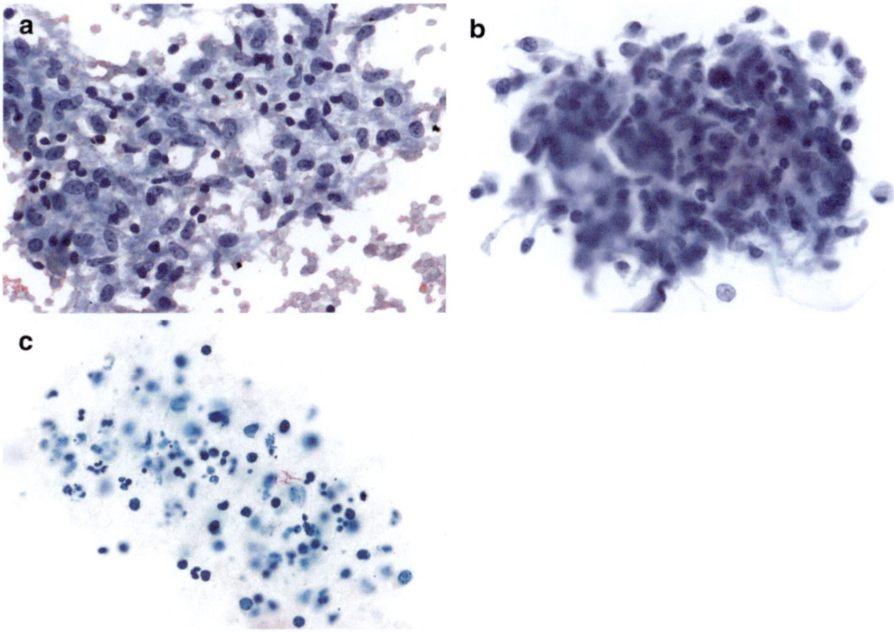

Fig. 9.1 Granulomatous inflammation. (**a**) This granuloma contains two types of cells: epithelioid histiocytes and lymphocytes. The histiocytes have round or curved hyphen-shaped nuclei that are more distinctive of histiocytes. Note how the small lymphocytes are scattered throughout the abundant foamy indistinct cytoplasm of the histiocytes (Pap-stained DS). (**b**) Although the granuloma in TP appears more three-dimensional, the histiocytic cells have similar nuclear and cytoplasmic features as the direct smear. The nuclear features are well preserved. Compare morphology of granuloma with SP preparation shown in Fig. 10.4b (TP). (**c**) Granulomas may be associated with autoimmune processes such as sarcoidosis as well as infectious organisms such as tuberculosis or yeast. Here, an acid-fast bacillus special stain highlights mycobacteria in a hot pink color. Additional LBP can be prepared for special stains or immunocytochemistry (AFB special stain on TP)

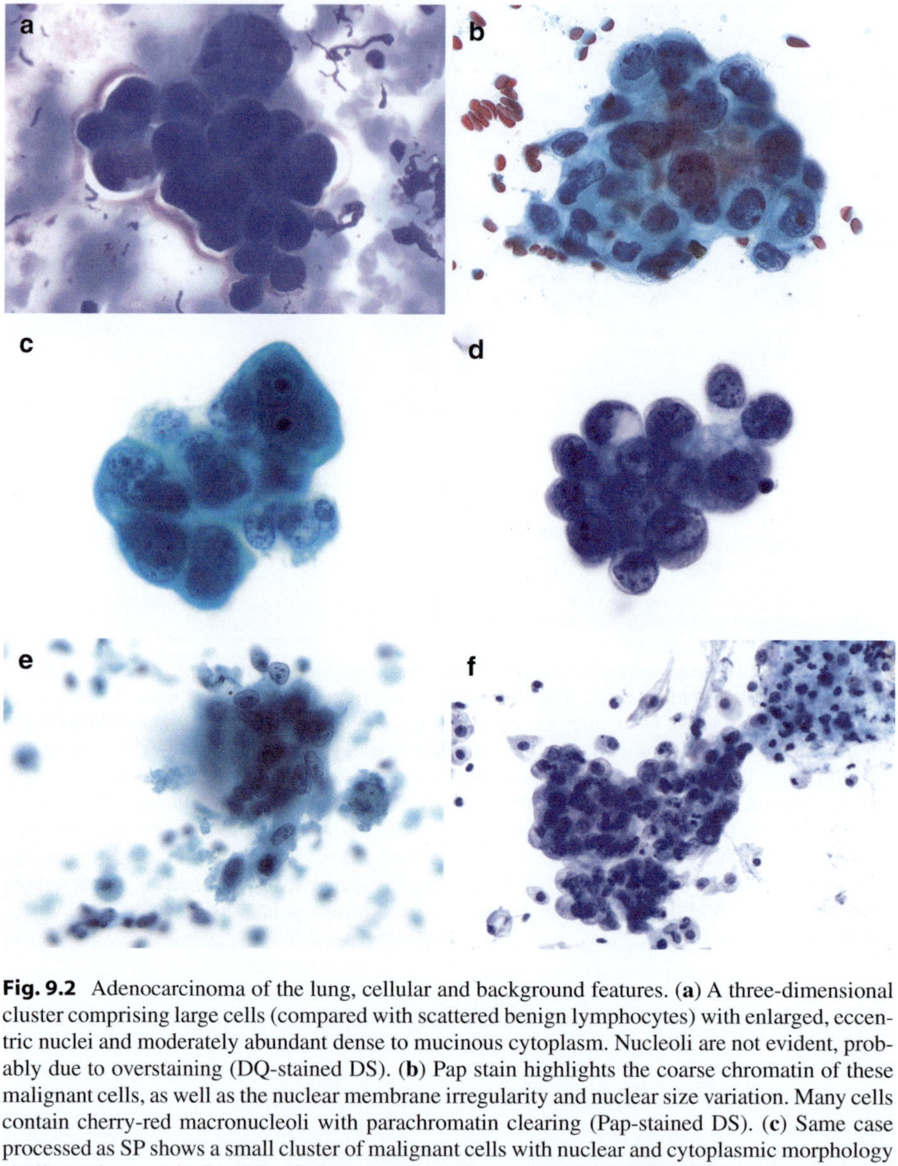

Fig. 9.2 Adenocarcinoma of the lung, cellular and background features. (**a**) A three-dimensional cluster comprising large cells (compared with scattered benign lymphocytes) with enlarged, eccentric nuclei and moderately abundant dense to mucinous cytoplasm. Nucleoli are not evident, probably due to overstaining (DQ-stained DS). (**b**) Pap stain highlights the coarse chromatin of these malignant cells, as well as the nuclear membrane irregularity and nuclear size variation. Many cells contain cherry-red macronucleoli with parachromatin clearing (Pap-stained DS). (**c**) Same case processed as SP shows a small cluster of malignant cells with nuclear and cytoplasmic morphology similar to that seen in (**b**) (SP). (**d**) Same case was also processed as a TP and shows a small cluster of malignant cells with nuclear morphology similar to that seen in (**b** and **c**). Cytoplasm appears more vacuolated and less dense than SP (TP). (**e**) Necrosis may occasionally be seen in adenocarcinomas of the lung. In SP, the malignant cell clusters are covered with necrotic debris that is also present free-floating in the background (SP). (**f**) Same case processed as TP shows clinging diathesis that does not obscure tumor cells (TP). Imura et al. [5] found that there were no significant cytomorphological differences between smears and LBP, except that some marginal nuclear swelling could be seen in smears. Moreover, sampling of viable tumor rather than necrotic areas is important for proper diagnosis and is one of the major advantages of ROSE. In TP, necrosis usually clumps or clings to tumor cells and does not obscure cell detail, as seen in image (**f**)

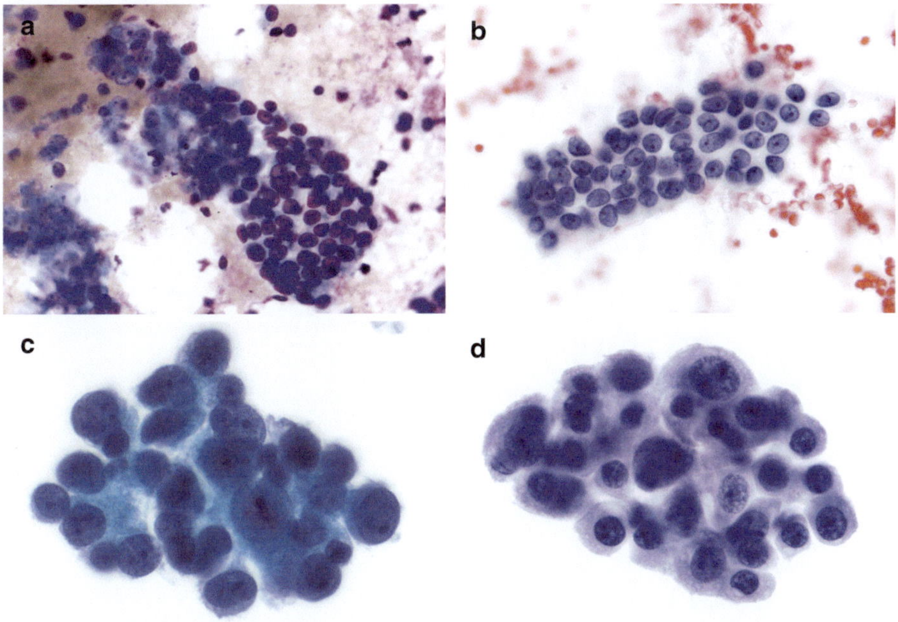

Fig. 9.3 Adenocarcinoma of the lung with lepidic growth pattern. Adenocarcinomas of the lung have various histological types based on architectural growth pattern. (**a**, **b**) The neoplastic cells appear monotonous in three-dimensional clusters (DQ) to flat sheets (Pap), with round nuclei showing minimal size variation and overlap, bland and granular chromatin, distinct nucleoli, mild nuclear membrane thickness, and irregularity. Cytoplasm is delicate and scant (**a**, DQ-stained DS and **b**, Pap-stained DS). (**c**, **d**) On TP the cells have similar nuclear and cytoplasmic morphology as on the smears, but the cells are more three-dimensional (TP). This pattern of growth was previously classified as "bronchoalveolar carcinoma" of the lung in which three-dimensional clusters of monotonous malignant cells with communal or scalloped borders and occasional intranuclear inclusions are seen [6]. (**e**) Compare the tumor cells with benign bronchial cells. Note the strips of single bronchial cells with clearly visualized terminal bar and cilia (TP). (**f**, **g**) Hyperchromatic and atypical nuclei can be seen lining airway spaces, which defines the "lepidic" growth pattern in this adenocarcinoma variant. Note benign bronchial cell lining on the right side of image **(g)** (H&E)

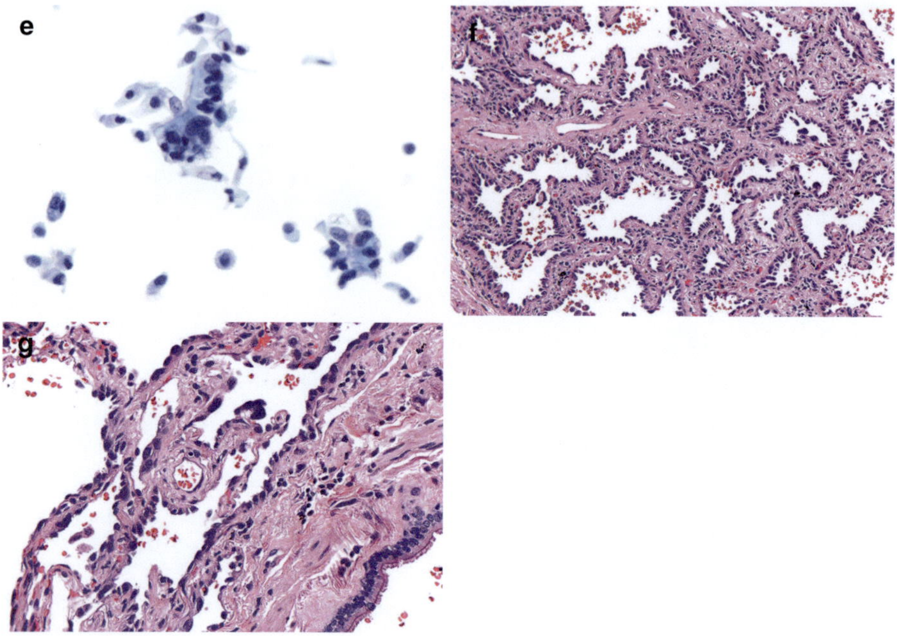

Fig. 9.3 (continued)

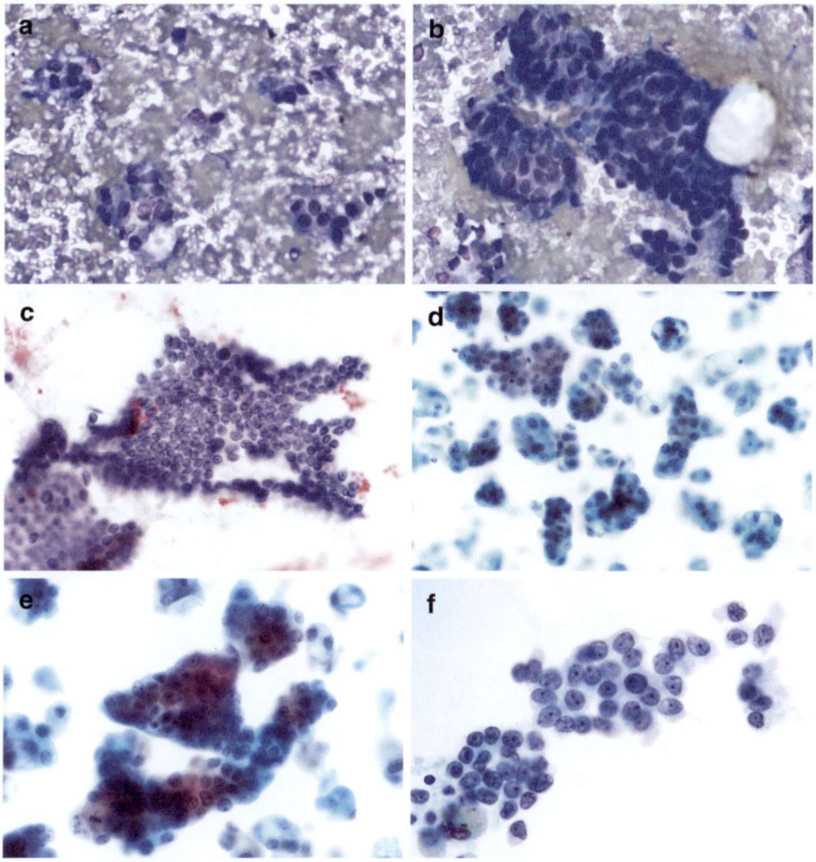

Fig. 9.4 Adenocarcinoma of the lung, acinar growth pattern. (**a**) Dispersed acinar arrangements of malignant cells with enlarged nuclei and elevated nucleus-to-cytoplasmic ratio (DQ-stained DS). (**b**) Larger fragments contain cells with enlarged and overlapping nuclei. Nuclei are arranged in rosettes and acinar formations (DQ-stained DS). (**c**) Monotonous round to elongated nuclei possess powdery chromatin, small nucleoli and minimal size, and nuclear border variation (Pap-stained DS). (**d**, **e**) Same case processed as SP showing dispersed islands of tumor cells in acinar formations in a clean background. Similar to the smears, the cells have small nucleoli and round, monotonous nuclei although the cytoplasm appears denser than the Pap-stained DS. Occasional vacuoles are seen within some cells (SP). (**f**, **g**) Same case was also processed as TP and shows distinct acinar arrangement. Compared to SP, the acinar formations are in flattened three-dimensional structures, and nuclear morphology is more clearly evident. Even in the larger fragment as seen in (**g**), the acinar formation is retained (TP). (**h**, **i**) Adenocarcinoma with acinar growth pattern shows irregularly shaped and variably sized glands with pleomorphic nuclei infiltrating fibrous tissue (H&E)

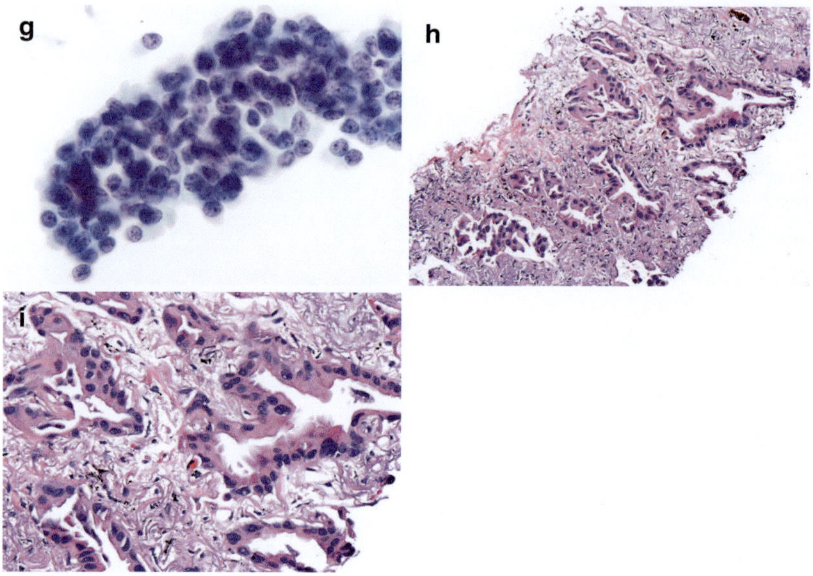

Fig. 9.4 (continued)

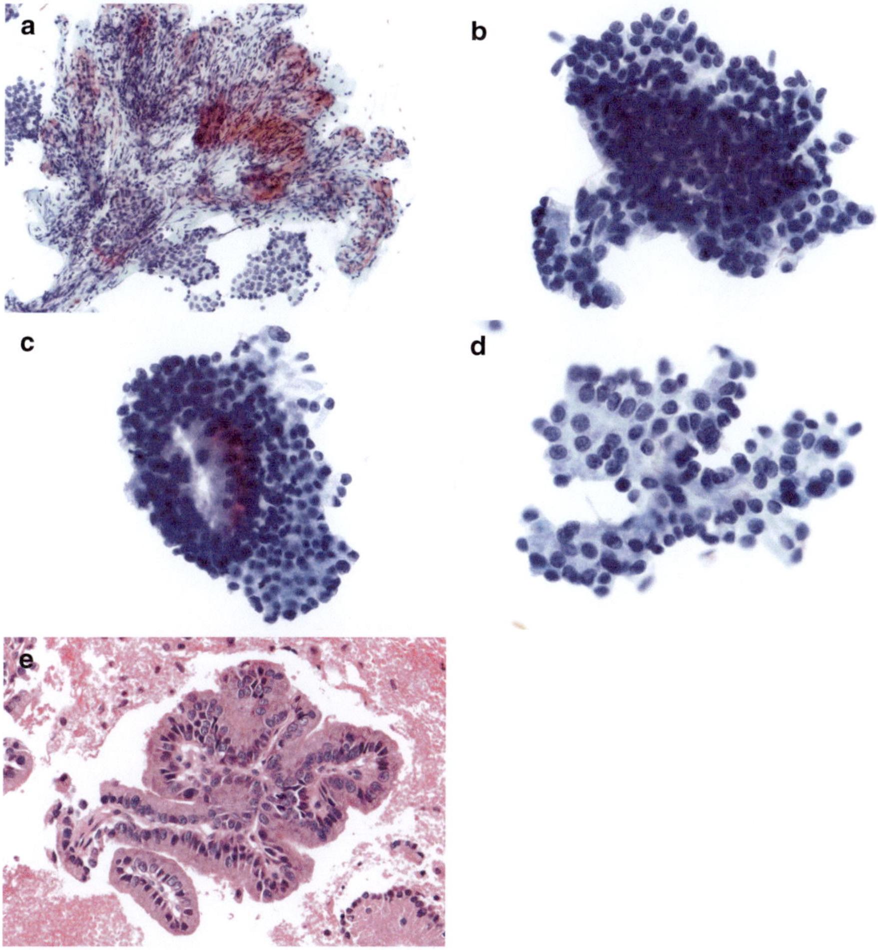

Fig. 9.5 Adenocarcinoma of the lung, papillary growth pattern. (**a**) From low magnification, the papillary architecture of the fragment can be appreciated (Pap-stained DS). (**b**) The LBP shows a papillary-like structure with small, overcrowded, and hyperchromatic malignant cells (TP). (**c**) Here the fragment takes an unusual shape, with a large gland attached to a sheet of tumor cells. The papillary growth pattern may have unusual structures such as this (TP). (**d**) Tumor cells forming complex small papillary projections. Nuclei are columnar with apical delicate cytoplasm (TP). (**e**) Cell block recapitulates the TP morphology (H&E)

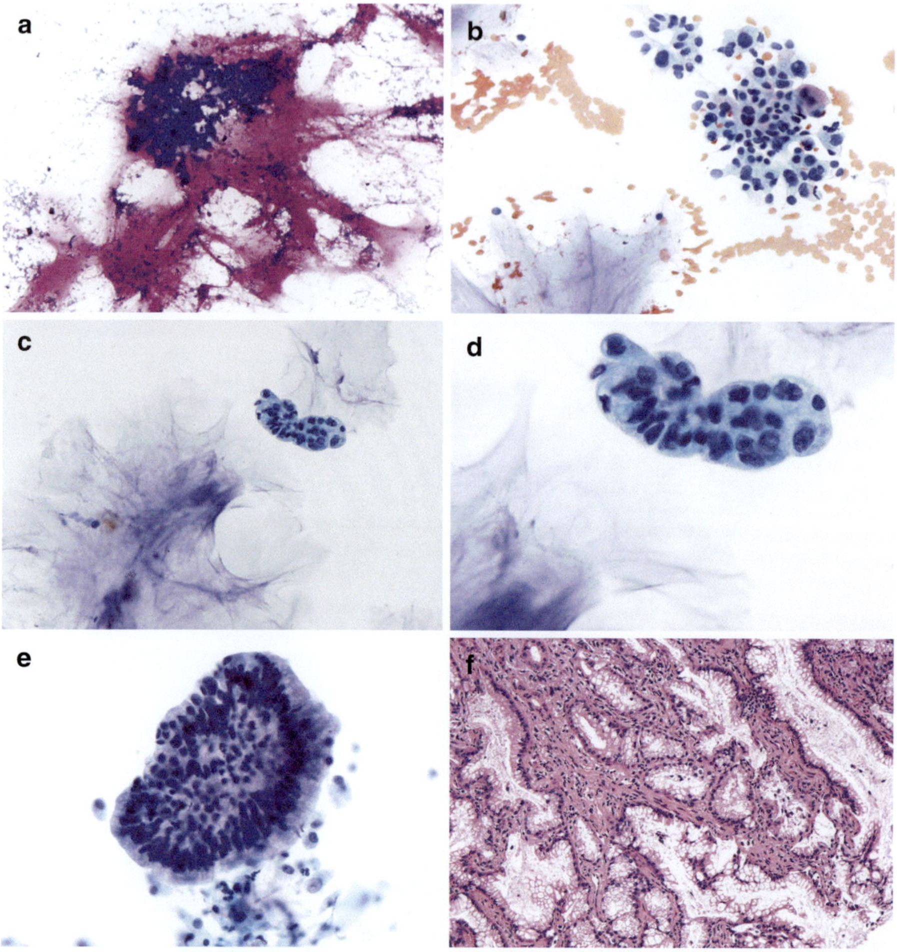

Fig. 9.6 Adenocarcinoma of the lung, mucinous type. (**a**) A small group of tumor cells is caught in a web of thick mucin, which stains magenta on this preparation. Such thick mucin is suspicious for a mucinous tumor (DQ-stained DS). (**b**) The alcohol-fixed smear shows better nuclear morphology. On the Pap stain, the thick mucin, more evident on the *left* side of the image, stains pale lavender. The nuclei are disorganized and variable in shape (Pap-stained DS). (**c**, **d**) Note similarity in mucin quality and cellular morphology in TP and Pap-stained DS. The carcinoma cells have dark and irregular nuclei, but maintain some amount of cytoplasm which reduces the amount of nuclear overlap. The lack of cilia suggests that this is not simply a fragment of reactive bronchial respiratory epithelial cells (TP). (**e**) While no mucin is present in the background, this fragment of cells contains apical mucin, and the nuclei generally remain basally located. Care should be taken to exclude mucinous metaplasia with reactive changes. In this case, many of the cells are dark and overlapping (TP). (**f**) Adenocarcinoma of the lung with mucinous features (H&E)

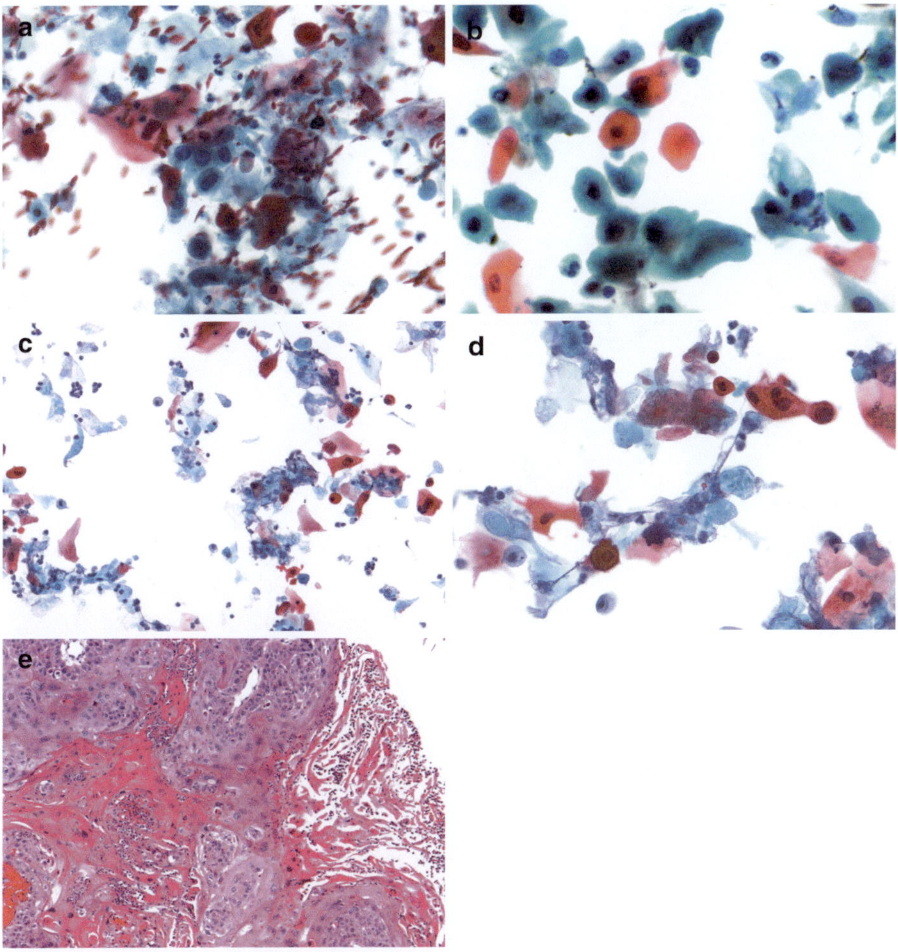

Fig. 9.7 Squamous cell carcinoma, keratinizing type. (**a**) The Pap stain is particularly useful for detecting keratinizing squamous cell carcinoma, as keratinized cytoplasm exhibits a distinctive orangeophilic appearance as seen here. Less keratinized cells have hyperchromatic, centrally located nuclei which are easily identified. Highly keratinized cells may have poorly stained nuclei, due to the poor penetration of the stain into the nucleus. This results in some cells looking deceptively bland (Pap-stained DS). (**b**) An SP preparation of the same case shows scattered keratinizing squamous cell carcinoma cells with irregular cellular shapes and nuclei so hyperchromatic that the chromatin pattern is not readily identifiable. The background shows scattered necrotic tumor cells (SP). (**c**, **d**) On TP, single keratinized squamous cells with pleomorphic nuclei and cytoplasm, a characteristic feature of keratinizing squamous cell carcinoma, are noted. Cellular necrosis and associated neutrophils are loosely attached to the malignant cells (TP). (**e**) On histology, the carcinoma has layers of keratinization and enlarged, irregular nuclei in many cells. Intervening pockets of necrosis can be seen (H&E)

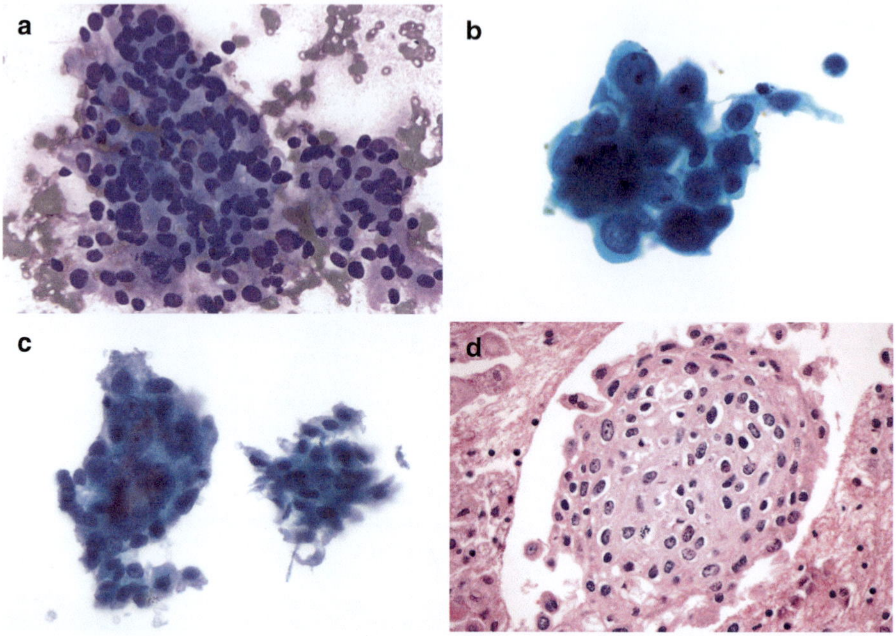

Fig. 9.8 Squamous cell carcinoma, nonkeratinizing type. (**a**) This large irregular sheet of malignant cells depicts loss of polarity, nuclear size variation and hyperchromasia, and dense cytoplasm. Poorly differentiated squamous cell carcinomas (without keratinization) can be difficult to distinguish from adenocarcinoma. However, the cells are usually dispersed as irregular sheets or dense clusters and do not exhibit glands formation. Because of low magnification, other nuclear features cannot be appreciated (DQ-stained DS). (**b**) Pap-stained SP, from a case of nonkeratinization squamous cell carcinoma, reveals a small and flat two-dimensional sheet of tumor cells with round nuclei, prominent nucleoli, thick nuclear membranes, and dense well-defined cytoplasm with rigid contours. Nucleus-to-cytoplasmic ratio is high. Careful evaluation shows possible intracellular bridges between the cells at the 5 and 11 o'clock positions, suggesting squamous differentiation. Immunostains for squamous cell nuclear markers, including, p40 and p63 were positive and those for adenocarcinoma including TTF-1 and napsin-A were negative (SP). (**c**) Example of nonkeratinization squamous cell carcinoma in TP. All the cytological features are as described for the SP. In addition, a small group of tumor cells show squamous cells with bipolar cytoplasm and elongated nuclei, a finding associated with invasive squamous cell carcinoma (TP). (**d**) Intracellular bridges are seen on this cell block preparation, most prominent at the 11 o'clock position. Nuclei are hyperchromatic and cytoplasm is moderately abundant and rigid (H&E). The use of cell block material in combination with TP has shown to be an effective method to obtain material for a cytological diagnosis and for immunohistochemical stain for p40 shows strong nuclear positivity supporting the diagnosis of squamous cell carcinoma. (**e**) p63 was also positive. (**f**) Positive CK5/CK6 immunostain highlights keratins found in the cytoplasm of most squamous cell carcinomas. CK5/CK6 and positivity for p40 and/or p63 are indicative of squamous cell carcinoma. (**g**) Histology of nonkeratinization squamous cell carcinoma (H&E).

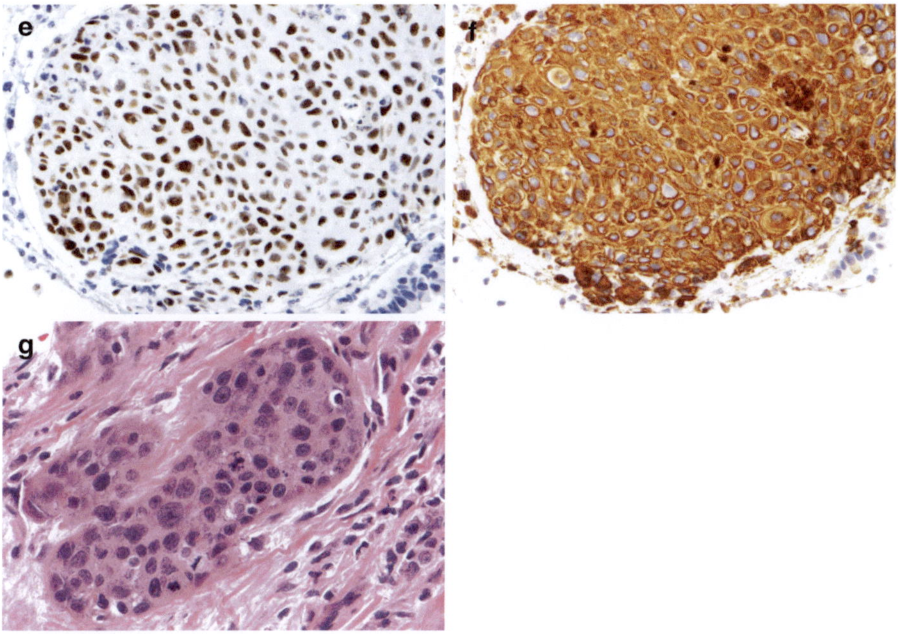

Fig. 9.8 (continued)

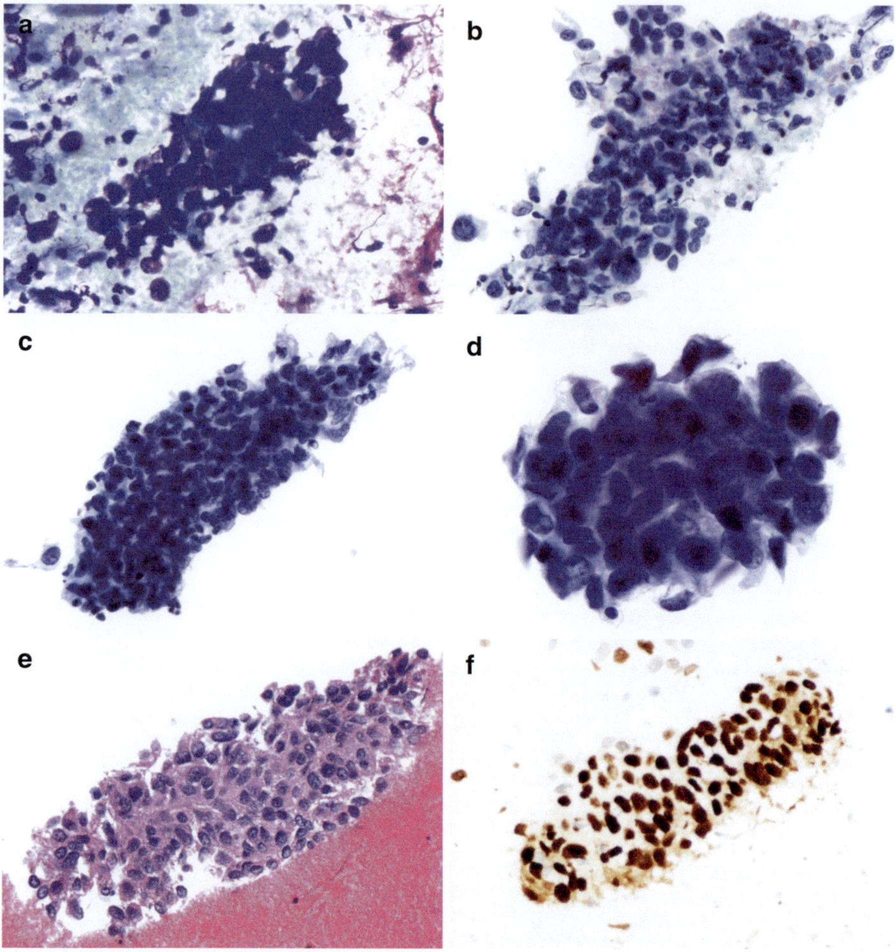

Fig. 9.9 Squamous cell carcinoma, basaloid type. (**a**) The malignant cells are large, pleomorphic, and basophilic and contain very little cytoplasm. There is pronounced nuclear overlap, resulting in a very basaloid appearance to the entire fragment. While the "basaloid" name of this variant results from its appearance on histology, this cytomorphologic pattern is similar to what is seen in other tumors with basaloid morphologies (DQ-stained DS). Differential diagnosis of basaloid squamous cell carcinoma includes small cell and adenocarcinoma. Pertinent immunostains to distinguish these three tumor types are usually performed. (**b**) Here the cytoplasm of cells is more identifiable and stains a pale color. The squamous features are hard to recognize, but the pleomorphic cells present in a fragment suggest a carcinoma. An area of necrosis is present attached to the right side of the fragment (Pap-stained DS). (**c**) A cohesive cluster of tumor cells in a TP preparation shows distinct basaloid squamous cell morphology. The nuclei are overlapping, small, dark (basaloid), and pleomorphic and contain very little cytoplasm (TP). (**d**) The above described nuclear features are much better appreciated in this image. Note the coarseness of the chromatin (TP). Background is clean and necrosis is not readily identifiable in both (**c** and **d**). (**e**) The CB preparation allows better visualization of the basaloid squamous cell carcinoma. Note the intercellular bridges (H&E). (**f**) Immunostain for p63 shows strong nuclear positivity (p63 immunostain)

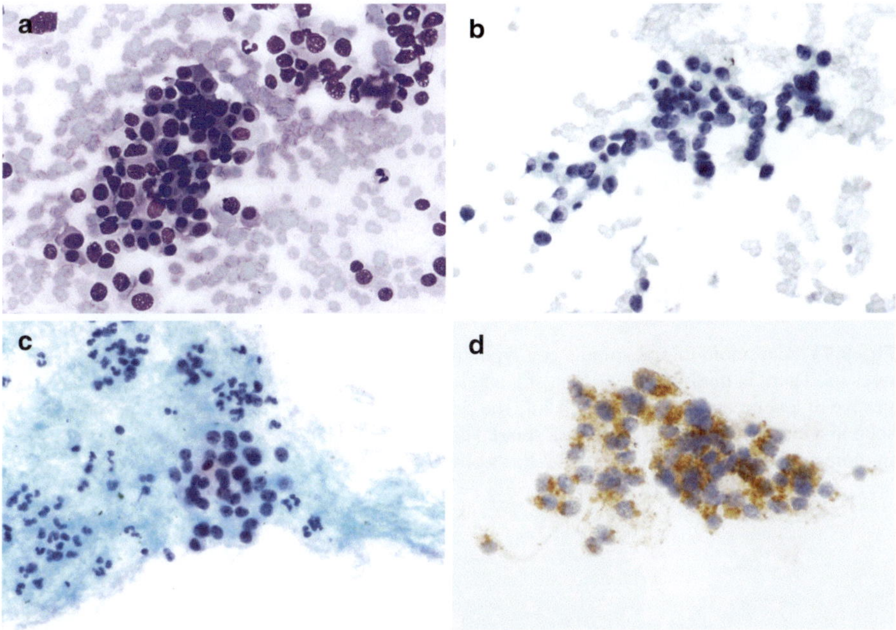

Fig. 9.10 Carcinoid tumor. (**a**) Monotonous-appearing epithelioid cells can be seen with round nuclei, homogenous chromatin, and moderate amount of eosinophilic cytoplasm. The cells have a plasmacytoid configuration and are present individually, in rosettes and in loosely cohesive groups (DQ-stained DS). (**b**) On the Pap stain, the cytoplasm is light colored and indistinct. The chromatin pattern can be better visualized as the "salt and pepper" quality often seen in neuroendocrine neoplasms. The nuclei are round and have minimal nuclear irregularities and occasional prominent nucleoli (Pap-stained DS). (**c**) A TP of the same case shows similar nuclear morphology as (**b**) the neoplastic cells are trapped in a clot intermixed with neutrophils, making the cytoplasm look indistinct (TP). (**d**) Carcinoid tumors are positive for neuroendocrine markers, such as synaptophysin and chromogranin (synaptophysin immunostain)

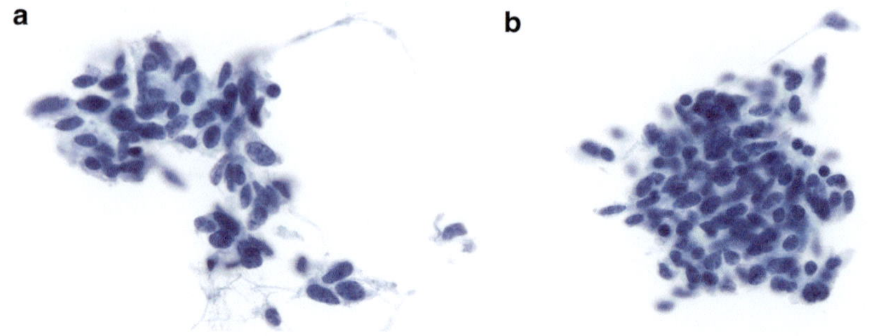

Fig. 9.11 Carcinoid tumor, spindle cell type. (**a**, **b**) The cells have a spindle shape and elongated nucleus. There is minimal pleomorphism, and the chromatin has a "salt and pepper" pattern suggestive of a neuroendocrine origin. While the cells form a small fragment, single cells can also be seen in a clean background (TP). In a spindle cell carcinoid, the plasmacytoid configuration seen in more epithelioid carcinoids may not be as obvious

Fig. 9.12 Small cell anaplastic carcinoma. (**a**) Small cells with little cytoplasm are present both in round and elongated shapes. Note the pleomorphism. Nuclei show molding, neuroendocrine features, and crush artifact (DQ-stained DS). (**b**) The Pap stain reveals the "salt and pepper" chromatin. Individual cells appear round, but in areas where the cells cluster, they gently "mold" against one another and become misshapen (Pap-stained DS). Some necrosis and scattered apoptotic bodies are also present in both preparations. (**c–e**) In TP of the same case, the neoplastic cells are dispersed throughout the field and also appear in loose clusters. Some clusters have necrotic debris attached, including ghostly cell forms representing cellular necrosis of the carcinoma. Some crush artifact is present in all three images. Nuclear molding is greatly reduced and appears more like overlap. Apoptosis is clearly visible in image (**e**) (**c–e**, TP). Studies have shown that TP results in decreased nuclear molding and nuclear smearing artifact, and more visible nucleoli and cytoplasm become more apparent due to immediate liquid fixation, [7]. (**f**) On the cell block preparation, only a thin crescent of cytoplasm can be appreciated. The nuclei are enlarged and the nuclear molding is more obvious. The chromatin pattern demonstrates the both coarse and granular combination seen in neuroendocrine neoplasms (H&E). (**g**) A synaptophysin stain is positive in the cytoplasm of the tumor cells, confirming neuroendocrine differentiation (synaptophysin immunostain). (**h**) Biopsy specimen shows prominent smearing artifact in which the chromatin is disrupted and smears as a string away from the cell, especially on the left-hand side of the field. In the better-preserved fragments, necrotic cells are present adjacent to viable cells, creating a checkerboard pattern (H&E)

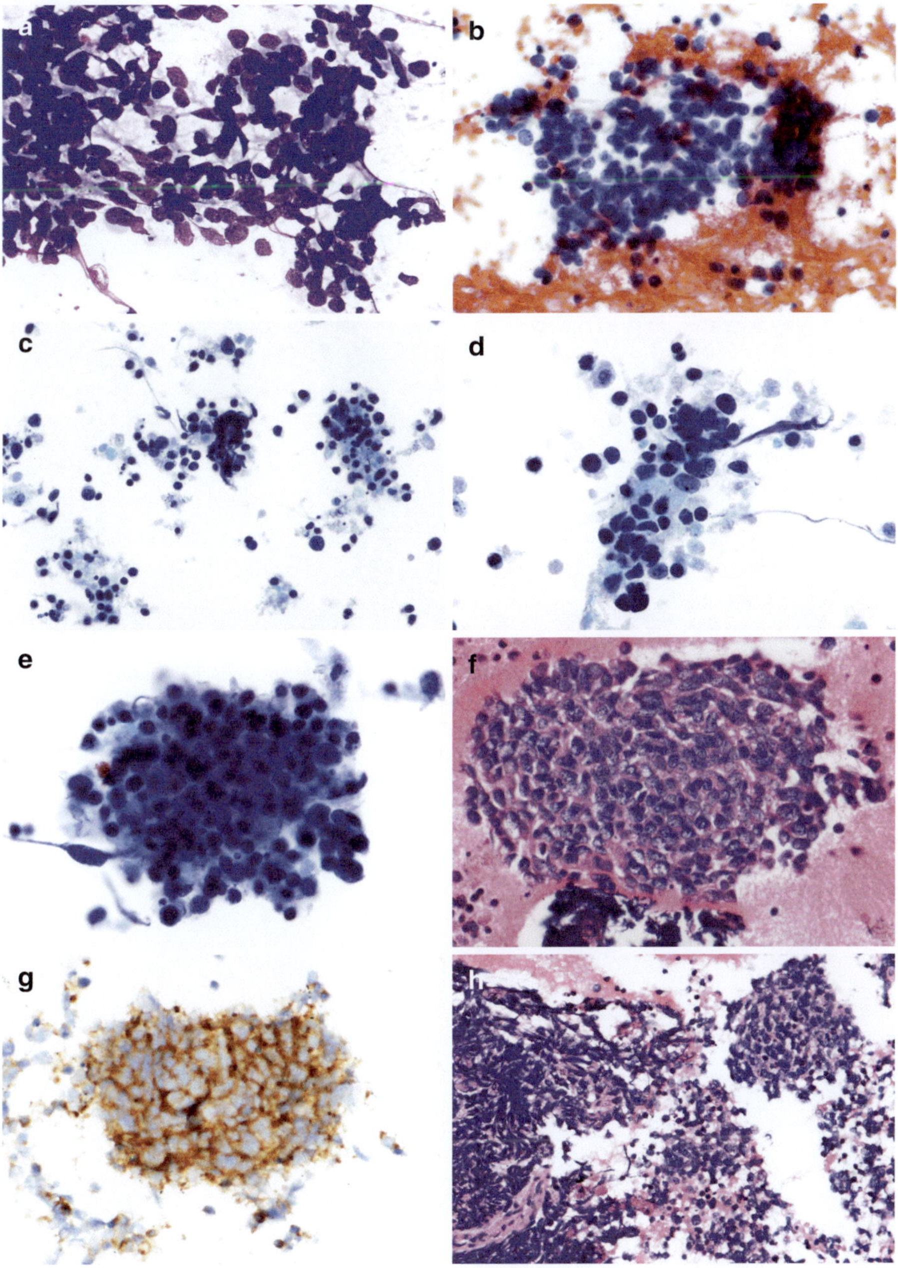
a
b
c
d
e
f
g
h

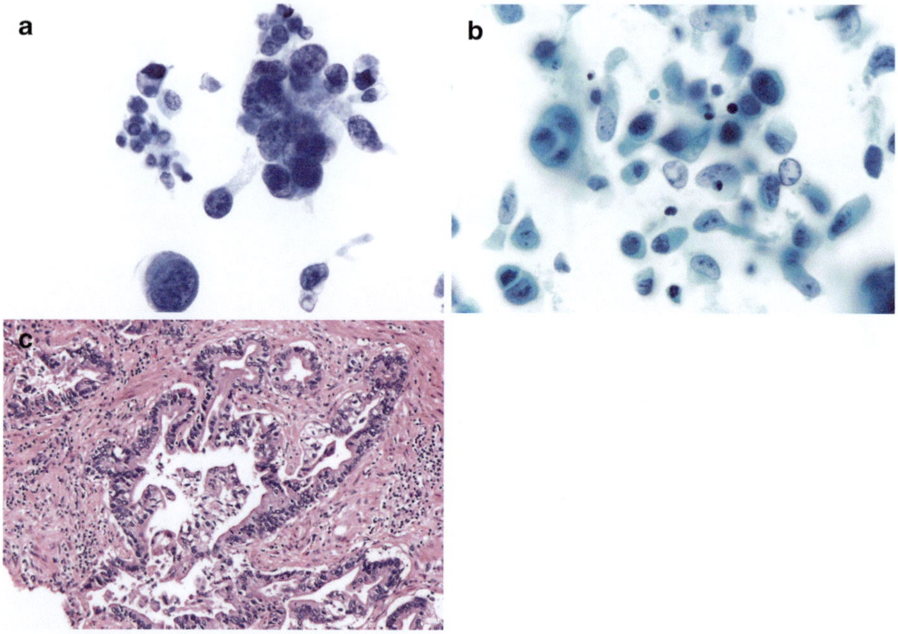

Fig. 9.13 Metastatic urothelial cell carcinoma. (**a**) The malignant cells have greatly elevated nucleus-to-cytoplasmic ratio, a coarse chromatin pattern, and dense cytoplasm which in some cell tapers giving the appearance of a "comet" which is a common cytomorphological feature of metastatic urothelial carcinoma. Immunostain for GATA3 was positive and TTF-1 and napsin-A were negative. The primary urothelial carcinoma of renal pelvis was also reviewed (TP) [8]. (**b**) More malignant "comet or cercariform cells" can be seen here; note the irregular nuclear borders and coarse chromatin quality. One cell appears to engulf another cell (*left-hand side* of the field), which can also be seen in urothelial carcinoma (SP). (**c**) Primary high-grade urothelial carcinoma of renal pelvis (H&E)

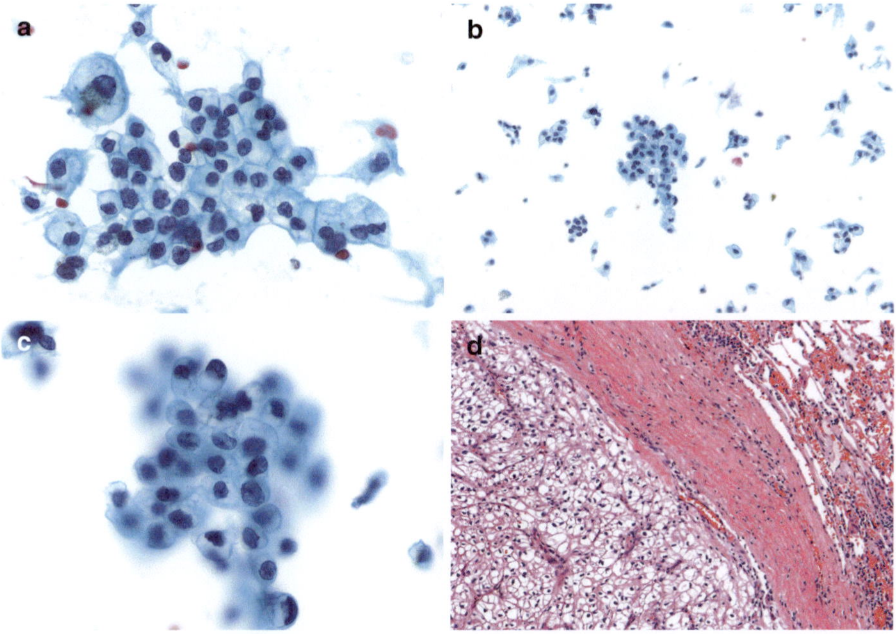

Fig. 9.14 Metastatic renal cell carcinoma. (**a**) The neoplastic cells have clear cytoplasm with distinct borders, and irregular nuclei with distinct nucleoli. Note the predominantly eccentric placement of the nuclei (Pap-stained DS). (**b**) TP shows high cellularity with both cellular clusters as well as dispersed single cells. The neoplastic cells have a similar morphology as in (**a**). While many carcinomas have high nucleus-to-cytoplasmic ratio, renal cell carcinoma often maintains its abundant cytoplasm even when high grade and metastatic (TP). (**c**) While the SP preparation maintains some three-dimensionality of this fragment, the cytology is similar to those in (**a**) and (**b**). The cytoplasm appears vacuolated (SP). (**d**) The resection specimen shows classic clear cell renal cell carcinoma histology, with nests of neoplastic cells with distinct cytoplasmic borders and clear cytoplasm separated by vessels (H&E)

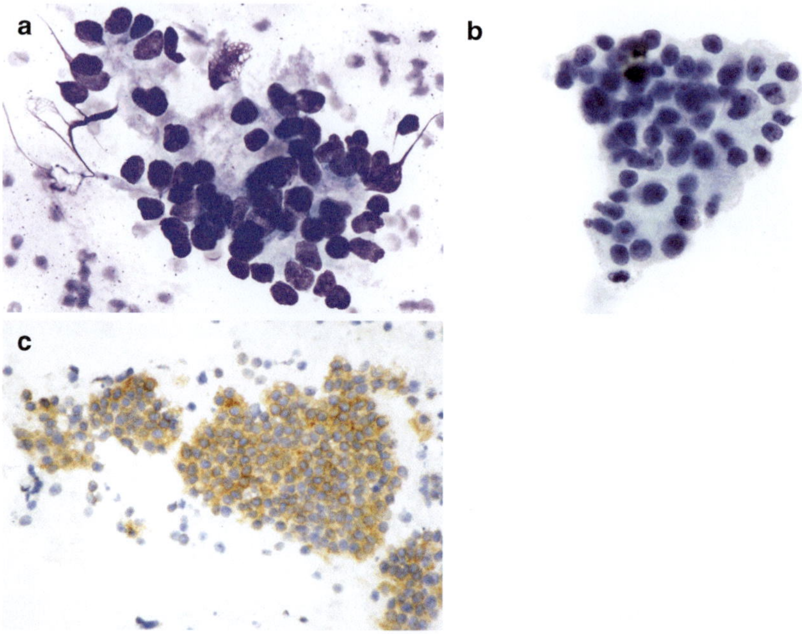

Fig. 9.15 Metastatic prostate adenocarcinoma. (**a**) Cells have abundant foamy cytoplasm and large, round nuclei. Nucleoli can be identified in some cells. Nuclei overlap in a rosette-like pattern, which recapitulates the cribriform architecture usually found in prostate adenocarcinoma (DQ-stained DS). **(b)** The Pap stain allows better visualization of the chromatin, and the prominent nucleoli can be in all the neoplastic cells; this is a common feature in prostate carcinoma. Note the very round nuclei with regular borders, which give the cells a deceivingly bland appearance and distinct rosette formation (TP). (**c**) Immunohistochemical stains can confirm the prostatic origin of this metastasis; commonly used stains include those for prostate-specific antigen (PSA), prostatic acid phosphatase (PSAP) shown here, and NKX3.1 (PSAP immunostain)

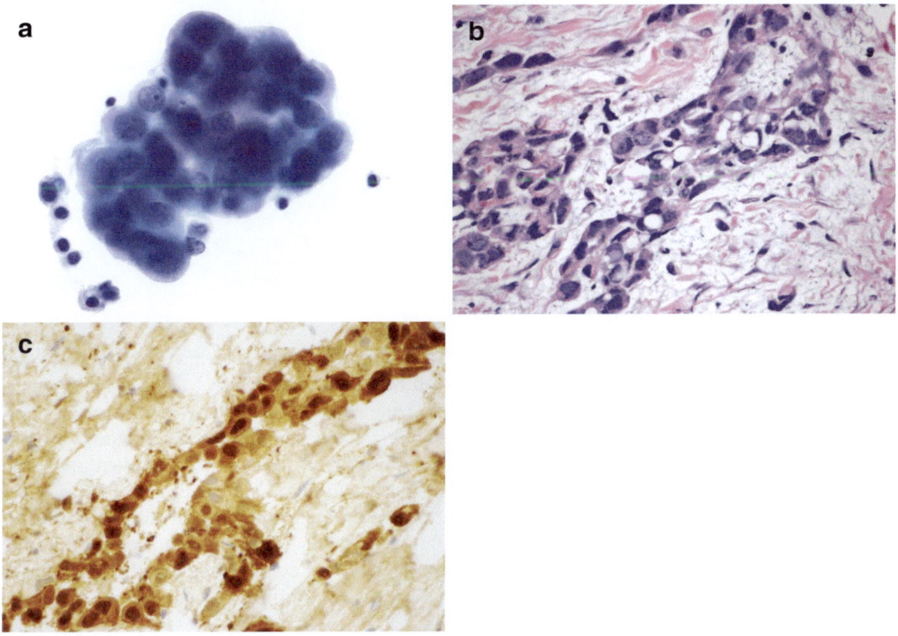

Fig. 9.16 Mesothelioma. Patient was a 68-year-old man with a pleural-based mass clinically suspicious for mesothelioma. (**a**) Large cells form a three-dimensional cluster with a hobnailed border. The cells and nuclei are round, and the nucleus-to-cytoplasmic ratio is elevated. The cytoplasm is distinct and basophilic. Some cells have coarse chromatin (TP). Distinguishing epithelioid mesotheliomas from adenocarcinoma can be challenging on cytomorphology alone. One study showed that enlarged nuclei were more frequently seen on TP as compared to DS, while DS demonstrated greater bubbly and vacuolated cytoplasm, cell-in-cell arrangements, and irregular nuclear contours than TP [9]. (**b**) Large cells with pleomorphic nuclei infiltrate fibrous tissue. A pleural-based location may be suggestive of mesothelioma as opposed to a carcinoma (H&E). (**c**) Nuclei and cytoplasm of the mesothelial cells are immunoreactive with calretinin immunostain. This immunostain is useful for proving a mesothelial origin, as most carcinomas are negative for calretinin (calretinin immunostain)

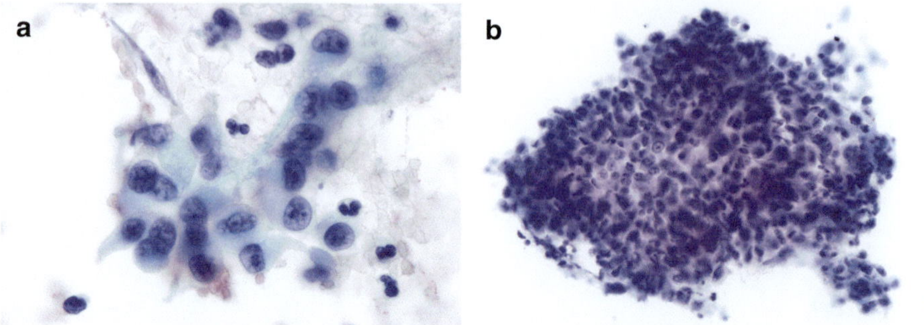

Fig. 9.17 Sarcomatoid mesothelioma. Mesothelioma can have many different morphologies and thus a high level of clinical suspicion, in comparison with radiologic studies, and the use of ancillary studies is critical. (**a**) Spindled cells with rounded nuclei giving the cells a sarcomatoid appearance that may mimic a spindle cell neoplasm including sarcomatoid renal cell carcinoma. Patient was 76-year-old shipyard worker with history of heavy smoking. Pleural plaques were noted on CT scan, and immunostains were positive for mesothelial cell markers and negative for epithelial cells (Pap-stained DS). (**b**) Similar findings can be seen on a TP of the same case. A cluster of malignant cells is seen, some with a spindled morphology. The nuclei are enlarged and overlapping with nucleoli, and a carcinoma may be suspected if one does not consider the possibility of mesothelioma and perform immunocytochemistry with clinical and radiological correlation (TP)

Suggested Reading

1. Konofaos P, Tomos P, Malagari K, Karakatsani A, Pavlopoulos D, Lachanas E, Flessas M, Kostakis A, Karakitsos P. The role of ThinPrep cytology in the investigation of lung tumors. Surg Oncol. 2006;15:173–8.
2. Minca EC, Lanigan CP, Reynolds JP, Wang Z, Ma PC, Cicenia J, Almeida FA, Pennell NA, Tubbs RR. ALK status testing in non-small-cell lung carcinoma by FISH on ThinPrep slides with cytology material. J Thorac Oncol. 2014;9:464–8.
3. Petriella D, Galetta D, Rubini V, Savino E, Paradiso A, Simone G, Tommasi S. Molecular profiling of thin-prep FNA samples in assisting clinical management of non-small-cell lung cancer. Mol Biotechnol. 2013;54:913–9.
4. Roy-Chowdhuri S, Aisner DL, Allen TC, Beasley MB, Borczuk A, Cagle PT, Capelozzi V, Dacic S, da Cunha SG, Hariri LP, Kerr KM, Lantuejoul S, Mino-Kenudson M, Moreira A, Raparia K, Rekhtman N, Sholl L, Thunnissen E, Tsao MS, Vivero M, Yatabe Y. Biomarker testing in lung carcinoma cytology specimens: a perspective from members of the pulmonary pathology society. Arch Pathol Lab Med. 2016;15 [Epub ahead of print]
5. Imura J, Abe K, Uchida Y, Shibata M, Tsunematsu K, Sathoh M, Tsuneyama K. Introduction and utility of liquid-based cytology on aspiration biopsy of peripheral nodular lesions of the lung. Oncology Letters. 2014;7:669–73.
6. Hoda R. Non-gynecologic cytology on liquid-based preparations: a morphologic review of facts and artifacts. Diagn Cytopathol. 2007;35:621–34.
7. Kim S, Owens CL. Analysis of ThinPrep cytology in establishing the diagnosis of small cell carcinoma of lung. Cancer. 2009;117:51–6.
8. Dey P, Amir T, Jogai S, Al JA. Fine-needle aspiration cytology of metastatic transitional cell carcinoma. Diagn Cytopathol. 2005;32:226–8.
9. Ylagan LR, Zhai J. The value of ThinPrep and cytospin preparation in pleural effusion cytological diagnosis of mesothelioma and adenocarcinoma. Diagn Cytopathol. 2005;32:137–44.
10. Wallace WA, Monaghan HM, Salter DM, Gibbons MA, Skwarski KM. Endobronchial ultrasound-guided fine-needle aspiration and liquid-based thin-layer cytology. J Clin Pathol. 2007;60:388–91.
11. Hou G, Yin Y, Wang W, Wang QY, Hu XJ, Kang J, Wu GP. Clinical impact of liquid based cytology test on diagnostic yields from transbronchial needle aspiration. Respirology. 2012;17:1225–8.
12. Leung CS, Chiu B, Bell V. Comparison of ThinPrep and conventional preparations: nongynecologic cytology evaluation. Diagn Cytopathol. 1997;16:368–71.

Cytologic Diagnosis of Lymphoproliferative Disorders by Morphology and Ancillary Techniques

10

Scott A. Ely

Introduction

The focus of the cytologist should be the well-being of the patient. The goal of a lymph node (LN) fine-needle aspiration (FNA) should be to provide a diagnosis and accompanying information sufficient to enable the clinical team to deliver the best care. The cornerstone of achieving this goal is good communication. To insure that the cytologist and clinical team are using the same language, current 2016 World Health Organization (WHO) terminology and diagnostic criteria must be used, and the report should state that WHO criteria are being used. For example, antiquated and/or vague terms should not be used (e.g., "malignant lymphoma" or "small cell lymphoma"). When possible, precise subtype diagnoses should be provided using exact WHO terminology. When the findings are insufficient for a WHO subtype diagnosis, a broader categorical diagnosis should be used, but as specific as possible (e.g., low-grade B-cell lymphoma). Also, when further studies are required, it should be stated explicitly (e.g., "For a more precise diagnosis, an excisional biopsy would be required.").

The surest and simplest way of making precise diagnoses and communicating adequate information well is using an algorithmic approach, detailed in this chapter for non-Hodgkin lymphoma (NHL) and Hodgkin lymphoma (HL). For an initial differential, considering lymphoma subtype relative incidence is helpful. Non-Hodgkin lymphomas (NHLs) generally are found in patients >60 years old. By contrast, HLs are commonly encountered in third-decade patients and can also be seen in adolescents. The difficulty of obtaining clinical information in some institutions, even with access to electronic medical records (EMR), is a well-known, ongoing obstacle in the path of practicing good medicine. In some cases, the diagnosis is clear based on the pathology findings alone. However, often communicating adequate information depends on the clinical context. In such cases, the importance of a concerted effort to obtain the necessary clinical information cannot be overemphasized.

R.S. Hoda et al., *Diagnostic Liquid-Based Cytology*,
DOI 10.1007/978-3-662-53905-7_10, © Springer-Verlag GmbH Germany 2017

Although therapy is beyond the scope of this chapter, and pathologists do not provide direct clinical care, it is important for the diagnostician to have some idea of the consequences of a subtype diagnosis and understand what ancillary testing data is required by the clinical team.

Polymerase Chain Reaction (PCR) and Fluorescence In Situ Hybridization (FISH)

If needed, an NHL diagnosis can be confirmed by PCR. Whereas PCR-positive (+) T-cell clones are common in reactive processes, finding a B-cell clone is rare in a reactive process and can be used to confirm B-NHL. FISH can be used for subtyping and also to establish clonality. PCR and FISH can be performed from any type of specimen [cells scraped off a smear slide, a section from a formalin-fixed, paraffin-embedded (FFPE) cell block, or liquid-based preparations (LBP)].

Reactive Lymphoid Infiltrates

The key to recognition of a reactive lymphoid infiltrate is cytologic variability. Because they are clonal, neoplastic processes are composed of monotonous populations. Even in lymphomas where neoplastic cells are the minority (e.g., HL), within the copious reactive background, there is a scattered population of monotonous, atypical cells. Aside from Hodgkin, finding numerous non-lymphoid cells (e.g., histiocytes and neutrophils) generally indicates a reactive process. Reactive infiltrates are composed mainly of small lymphocytes with scant cytoplasm, round nuclei, and clumped chromatin. Although it is not possible to distinguish a B cell from a T cell with certainty based on cytologic features alone, a good guess is possible and should be attempted. Specifically, because T cells are in the majority in reactive infiltrates and because they typically have identifiable features, after cytologic variability has been established, a rough assessment of the proportion of T cells should be made. The keys to cytologic distinction of small T cells:

Reactive T-Cell Features

- Small size
- Extremely dark, featureless chromatin

Infectious Processes, Including Granulomatous Inflammation

Because the identification of infectious organisms can only be made by cultures and associated techniques, it is critical whenever an infectious process is suspected, to perform a needle pass to obtain and send material to the microbiology lab. This mainly applies to finding of a granulomatous process. If a specimen shows granulomas and material was not sent to microbiology and if acid-fast and GMS stains were negative, it is important to communicate to the clinical team that, although special stains have high specificity, their sensitivity for detecting organisms is notoriously low.

B-Cell Lymphoma

In contrast to reactive processes, which are composed mainly of T cells, most lymphomas are B-NHL, which are composed mainly of B cells (T cells are present in the background, but fewer in number). The suspected diagnosis of lymphoma by cytologic morphology must be confirmed by immunohistochemistry (IHC) or flow cytometry [multiparameter flow cytometry (MFC)]. Although there are hundreds of CD (cluster of differentiation) and other antibodies for clinical use, a relatively small number are necessary for lymphoma diagnosis.

There are a few standard ways to establish the diagnosis of B-NHL in a cytology case with convincing clinical and cytologic features.

Low-Grade (LG) B-Cell Lymphoma

Monotonous small lymphocytes, absent necrosis

- + B-cell antigen expression by immunohistochemistry (IHC)
- ± confirmatory aberrant antigen expression
- + positive MFC
- + equivocal or suggestive MFC and/or IHC + positive PCR
- + equivocal or suggestive MFC and/or IHC + positive FISH

High-Grade (HG) B-Cell Lymphoma

Monotonous or scattered, markedly atypical large transformed lymphocytes, often with background necrosis

- + B-cell antigen expression by IHC
- + positive MFC
- + equivocal or suggestive MFC and/or IHC + positive PCR
- + equivocal or suggestive MFC and/or IHC + positive FISH

Once the diagnosis of B-NHL is made based on these findings, a subtype diagnosis can be established by application of an algorithm.

Plasmacytoid Differentiation as a Tool to Establish the Diagnosis of B-NHL

Most B cells express only surface immunoglobulin (sIg). Expression of cytoplasmic Ig (cIg) denotes plasmacytoid differentiation. Reactive populations contain polytypic plasmacytoid cells (i.e., cIgκ:cIgλ = ~2:1). IHC or in situ hybridization (ISH) for Ig light chains or light chain RNA, respectively, can be used to detect a monotypic plasmacytoid subpopulation, which establishes the diagnosis of a lymphoproliferative disorder (Fig. 10.8). It can also be used for subtyping. Plasmacytoid differentiation is seen in 100 % of LPL/WM, 30 % of MZL, 10 % of CLL, 5 % of FL, and never in MCL.

Use of CD43 IHC to Establish the Diagnosis of B-NHL

CD43 is normally expressed by T cells, myeloid cells, and plasma cells (PC). Expression by B cells is aberrant. Finding a large population of B cells with CD43 expression usually confirms the diagnosis of B-NHL. CD43 is expressed by 80–90 % of mantle cell lymphoma (MCL) and chronic lymphocytic lymphoma (CLL), 30 % of marginal zone lymphoma (MZL), and almost never in follicular lymphoma (FL). In cases with clear coexpression on a large population of monotonous B cells, in combination with compelling cytologic and clinical features, CD43 can be used to confirm the diagnosis of B-NHL. If expression is weak, partial, or less clear, diagnostic confirmation by other ancillary tests is required.

Low-Grade (LG) B-Cell Lymphoma

Chronic Lymphocytic Leukemia (CLL) and Mantle Cell Lymphoma (MCL)

For diagnostic purposes, CLL and MCL are best considered together due to cytologic similarity and coexpression of CD5. Aside from rare, CD5+ marginal zone lymphoma and even rarer CD5+ de novo, diffuse large B-cell lymphoma (DLBCL), expression of CD5 is only seen in CLL and MCL. Although reactive CD5+ B-cell subsets are unusual in lymph nodes, they are sometimes encountered in body fluid specimens. MCL is best distinguished from CLL by expression of cyclin D1, which, among lymphoproliferative disorders (LPD), is only seen in MCL, in hairy cell leukemia (HCL; below), and in some plasma cell neoplasms (PCN; below). Although there is published data suggesting the existence of cyclin D1-negative (−) MCL, it is not well established. SOX11 expression, for example, is not sufficient for the diagnosis of MCL.

Though MCL shares some cytologic features with CLL, MCL cell size is more commonly intermediate, and rather than round nuclei, the contours are commonly irregular. A good rule of thumb: if the cytology is difficult to categorize as low grade (LG) or high grade (HG), consider MCL. It should be noted that, whereas most lymphomas are limited to lymphoid organs and marrow, CLL is leukemic and, therefore, typically found to greater or lesser extent in any specimen, from any site, in a CLL patient. When a minor population of CLL cells is detected in a CSF or fluid specimen, such as by MFC, the report should be worded with caution, "Although there are X % CLL cells in the specimen, suggesting the possibility of CLL in this site, the differential includes peripheral blood contamination."

Follicular Lymphoma (FL)

Whereas lymphomas show a monotonous population, FL is the exception to the rule. FL is characteristically composed of cells of variable size. As such, it often is difficult to distinguish from a reactive infiltrate. IHC should show a majority CD5 (−) B cells with coexpression of BCL6 (± CD10). MFC should show sIg light chain restriction, although some FL cases are sIg(−). If needed, PCR can be performed for B-cell clonality and to establish a BCL2 gene translocation. Also, FISH can be performed for t(14;18), which is positive in most cases.

Marginal Zone Lymphoma (MZL)

The three WHO types of MZL are characteristically restricted to certain sites: extranodal MZL (E-MZL), nodal MZL (N-MZL), and splenic MZL (S-MZL, which involves spleen, marrow, ± peripheral blood). The characteristic cytology is monocytoid, although some cases have less distinctive, generic small lymphocyte morphology. Because MZL does not typically show aberrant or distinctive antigen coexpression, unless the diagnosis is confirmed by CD43 or plasmacytoid differentiation (see above sections), MFC, PCR, or FISH are required.

High-Grade B-Cell Lymphoma (HG B-NHL)

Aside from some cases of ALL (discussed below) with relatively small cells, high-grade lymphomas have large or at least intermediate-sized nuclei. Whereas low-grade lymphomas have clumped chromatin, high-grade lymphomas have disbursed, open chromatin, often with visible nucleoli. Necrosis is common in the background of HG and rare in LG lymphomas. When HG B-NHL is suspected based on morphology and clinical information, the diagnosis can be confirmed by IHC or MFC. An algorithm can be applied for subtyping and guidance in ordering ancillary studies.

Diffuse Large B-Cell Lymphoma (DLBCL)

DLBCL is the most common NHL. Patients typically present with lymphadenopathy. FNAs typically show a monotonous population of large lymphocytes with disbursed chromatin and prominent nucleoli, often in a necrotic background. Diagnosis can be confirmed by IHC or MFC. Ancillary testing should always be performed to exclude "double-hit lymphoma" (below). Although much has been written about the separation of DLBCL into germinal center (GC) and non-germinal center/activated B-cell (ABC) types, this distinction does not affect therapeutic decisions and has a currently questionable effect on prognosis. However, in many centers the Hans algorithm is still performed.

Double-Hit and Double-Protein Lymphomas

In most centers, cases otherwise categorized as DLBCL but shown to harbor both BCL2 and cMYC gene translocations are designated as double-hit lymphoma. These patients receive chemotherapeutic regimens that are different from DLBCL. Although the designation "double-protein lymphoma" is less well established, treatment of such patients differs from DLBCL in many centers. In summary, in light of these considerations, in high-grade B-NHL with Ki67 > 80 %, IHC and FISH should be performed for MYC and BCL2.

Burkitt Lymphoma (BL)

Burkitt lymphoma is relatively rare and characteristically occurs in children and AIDS patients. Unlike other lymphomas, BL is typically extranodal (e.g., intra-abdominal, involving the bowel wall or ovaries; breast) and/or leukemic. Outside of these isolated circumstances, the diagnosis of BL should be approached with caution. The immunophenotype also is characteristic: CD10+, BCL6+, BCL2−, and Ki67 > 90 %. In endemic (e.g., Africa) and in AIDS patients, it is EBV+. FISH for t(8;14) should be performed but is not strictly necessary.

Mantle Cell Lymphoma (MCL)

Though commonly cytologically low grade, MCL cells often are intermediate in size, difficult to categorize as LG or HG. Also, some patients undergo a "blastoid" transformation, in which the cells resemble lymphoblasts or, rarely, large cell lymphoma. Because MCL therapy is the same, regardless of cytologic features, it should be considered even in cases with HG cytologic features. Because coexpression of CD5 and cyclin D1 is unique and consistent, IHC can confirm the diagnosis.

Acute Lymphoblastic Lymphoma, ALL (B or T Lymphoblastic Leukemia/Lymphoma)

"ALL," which originally denoted acute lymphoblastic leukemia, is a term so prevalently used, it is somewhat of an exception to the rule regarding adherence to WHO terminology. Because ALL mainly presents as a fulminant leukemia, it is rare to receive a "r/o ALL" specimen in an undiagnosed patient. When suspected by morphology, the diagnosis can be confirmed by the typical coexpression of TdT and CD34 (with PAX5+ and/or CD79a+ in B ALL; CD7+ and/or cytoplasmic CD3+ in

T ALL). TdT is expressed by >90 % of cases and is particularly helpful because it is almost unique to ALL and can distinguish it from neuroendocrine tumors that, like ALL, express CD99.

T-Cell Lymphomas

The most common T-NHL are mycosis fungoides/Sezary syndrome (MF/SS), anaplastic large cell lymphoma (ALCL), and T-NHL, not otherwise specified (NOS). Diagnosis is fairly straightforward. PCR should be used to confirm a diagnosis by demonstrating a monoclonal T-cell population, but is not a primary means of establishing a diagnosis because T-cell clones are common in reactive processes. FISH is generally not necessary in T-NHL.

Lymphomas Composed Mainly of Reactive Cells, with a Minority of Malignant Cells

Hodgkin Lymphoma (HL), T-Cell/Histiocyte-Rich Large B-Cell Lymphoma (THRLBCL), EBV+ DLBCL, NOS*

Although these three lymphomas have vastly different clinical features and therapy, they all occur in lymph nodes and have similar morphologic features.

Plasma Cell Neoplasms (PCN), Including Multiple Myeloma (MM)

Diagnostic criteria for MM are made and updated by the International Myeloma Working Group (IMWG). These updates are more frequent than those of the WHO. When issued, WHO criteria comprise IMWG criteria at that point in time. Although the criteria are interchangeable, while the IMWG and nearly all literature use the term MM, the WHO instead uses the term plasma cell myeloma (PCM).

The simplest way to establish the diagnosis of a PCN is demonstration of aberrant antigen expression. Because the most common form of aberrant antigen expression is weak/partial expression of cyclin D1 (seen in hyperdiploidy), multiplex IHC (mIHC for CD138/cyclin D1) is the simplest, surest first-line study. mIHC with various other combinations can be used to detect PCN with other cytogenetic features (e.g., cyclin D3 is aberrantly expressed with t(6;14)). Of the various ways to prognosticate in MM, the most cost effective is to assess the plasma cell-specific proliferation index (PCPI), by one of the two validated methods.

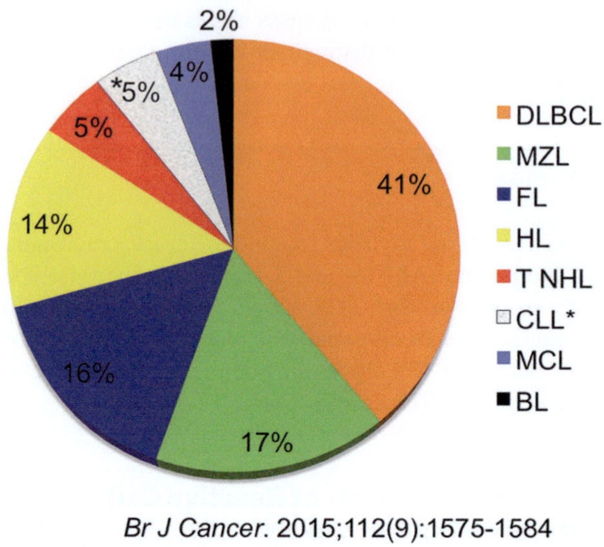

Fig. 10.1 Relative proportional incidence of lymphomas. DLBCL (diffuse large B-cell lymphoma), MZL (marginal zone lymphoma), FL (follicular lymphoma), HL (Hodgkin lymphoma), T-NHL (T-cell non-Hodgkin lymphoma), CLL (chronic lymphocytic leukemia), MCL (mantle cell lymphoma), and BL (Burkitt lymphoma). These relative incidences of lymphoma subtypes, from Britain in 2015, are paralleled by data from other countries, including the USA. *The incidence of CLL is difficult to assess because, whereas data for other types are clear, registries divide CLL patients between categorization as lymphoma and as leukemia

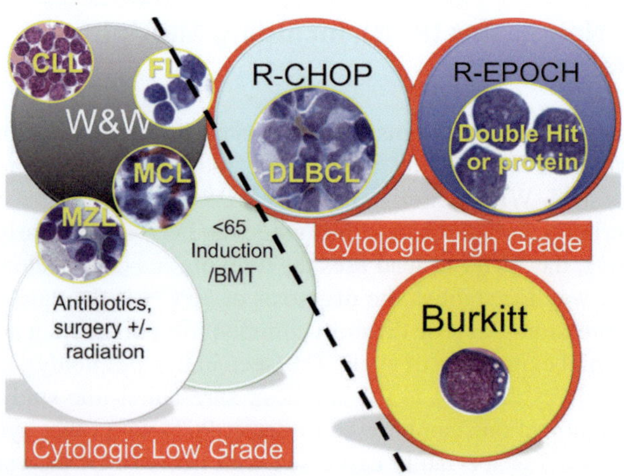

Fig. 10.2 Rough overview of therapeutic regimens. A rough overview of therapeutic approaches (large circles) divides (dashed line) lymphomas into cytologic low-grade (LG) and high-grade (HG) categories. For LG (CLL, FL, MZL, and MCL), the approach in most centers is "watch and wait" (W&W) or expectant management, because most are incurable. Exceptions include extranodal MZL, which is curable and usually receives antibiotics and/or local therapy, and MCL, which, in some centers, is treated with *bone marrow* transplantation (BMT) in younger patients. By contrast, diffuse large B-cell lymphoma (DLBCL) is treated with a multi-agent, outpatient chemo regimen (RCHOP), and double-hit or double-protein lymphomas are treated with more intensive, in patient therapy (R-EPOCH). For BL, there are numerous therapeutic options, all involving chemotherapy

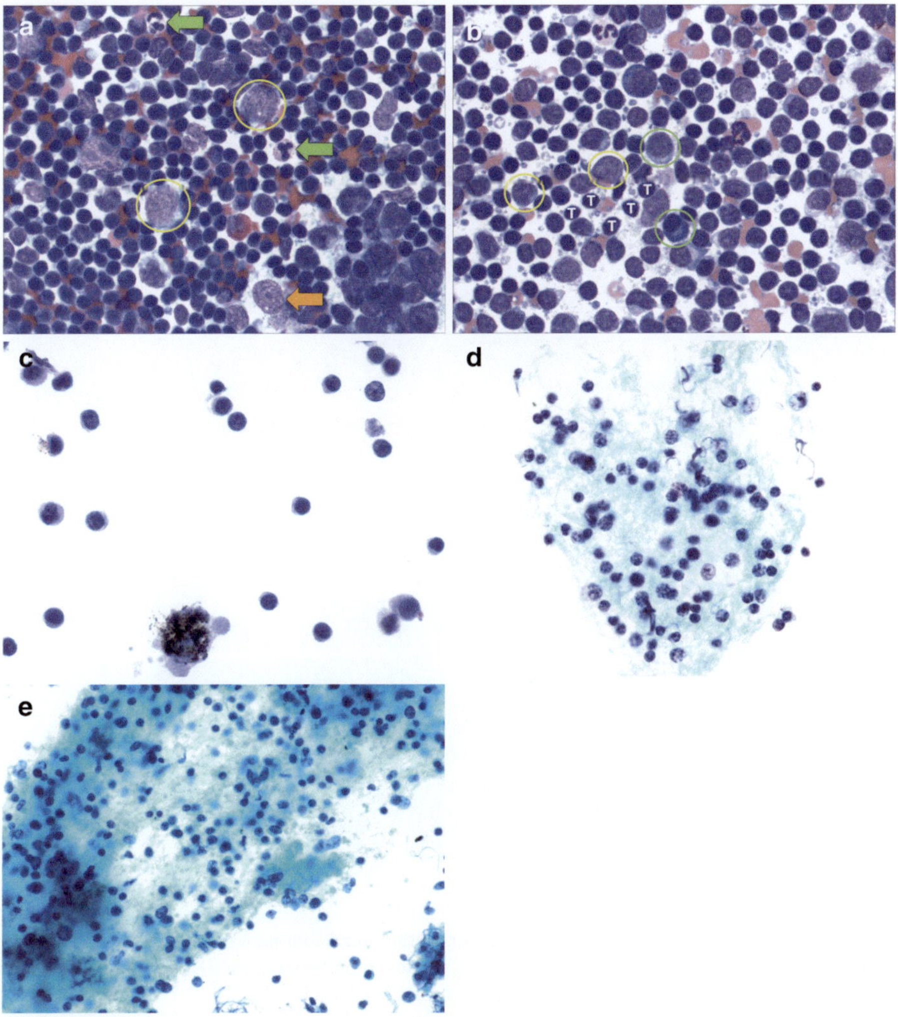

Fig. 10.3 Reactive lymph node infiltrate. (**a**) FNA smear from a reactive lymph node typically shows broad cytologic variability. Although composed mainly of small lymphocytes, the infiltrate also contains scattered, large transformed lymphocytes (*yellow circles*) as well as neutrophils (*green arrows*) and histiocytes (*orange arrow*). Neither cytologic monotony nor atypia is prominent in a reactive infiltrate (DQ-stained DS). (**b**) FNA smear shows a strikingly polymorphous population. As is typical of a reactive infiltrate, it is composed mainly of T cells ("T"), which are very small with round nuclei and strikingly dark, featureless chromatin. They are admixed with intermediate-sized lymphocytes (*orange circles*) and occasional plasmacytoid cells (*green circles*) with moderate to abundant blue cytoplasm, eccentric nuclei, and juxtanuclear Golgi (DQ-stained DS). **(c-e)** Reactive lymph node on TP and SP shows the same features as noted in DQ stain. Note the tingible body macrophage at the bottom of the image (**c**) (**c** and **d**, TP; **e**, SP)

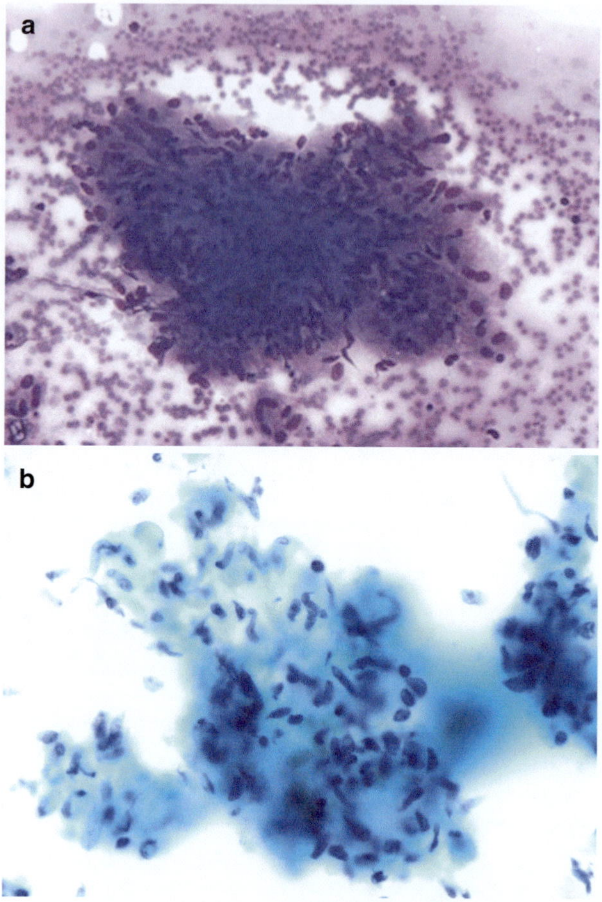

Fig. 10.4 Granuloma. (**a**) and (**b**) Non-caseating granuloma in a patient with sarcoidosis. Note lack of necrosis and mixed, variably-shaped histiocytes, more evident on SP, and lymphoid cells. Overall features are similar in both preparations (**a**, DQ-stained DS; **b**, SP)

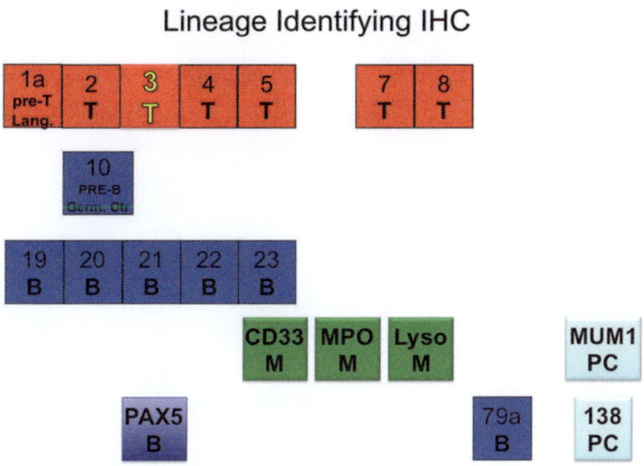

Fig. 10.5 Lineage identifying immunohistochemistry (IHC). Pan-T-cell antigens that should be expressed by all peripheral T cells and include CD2, CD3, CD5, and CD7; expression of CD3 defines T-cell lineage; also, all peripheral T cells should express CD4 or CD8 (not both, not neither). B-cell antigens are those that express CD20, but the best for IHC are CD19, CD79a, and the nuclear antigen, PAX5. Also note markers for myeloid (M) and plasma cells (PC)

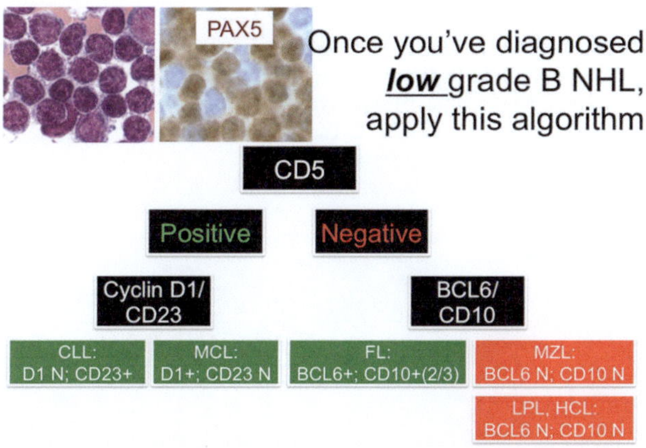

Fig. 10.6 IHC algorithm, low-grade B-NHL. Once the diagnosis of low-grade B-NHL is established by morphology and surface immunoglobulin light chain expression, or by morphology and immunostaining, subtyping begins with assessment of CD5. CD5 is positive in CLL and MCL. If CD5 is positive, the next step is assessment of cyclin D1, a nuclear antigen, aberrantly expressed in MCL, but negative in CLL. For technical reasons, cyclin D1 expression cannot be assessed by multiparameter flow cytometry (MFC). Though cyclin D1 assessment is necessary to exclude MCL, another means of distinguishing CLL from MCL is CD23, which is typically positive in CLL but negative in MCL. If the low-grade B-NHL is CD5 negative (−), use BCL6 to distinguish between FL, which is always BCL6+, and MZL, which is always BCL6−

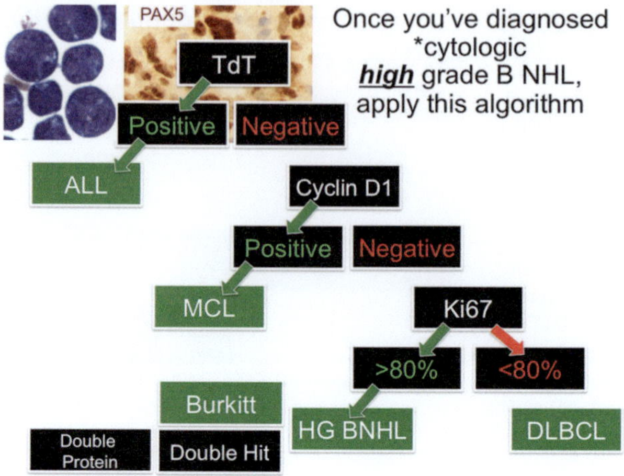

Fig. 10.7 IHC algorithm, high-grade B-NHL. If your cytologic assessment is high-grade lymphoma and you confirm B-cell origin by IHC or MFC, the algorithm begins with TdT, which is positive in most acute lymphoblastic lymphoma (ALL) and rarely in any other cancer. If TdT is negative, use cyclin D1 to exclude MCL. Although MCL is usually cytologically low grade, some cases have intermediate cytologic features and others represent blastoid transformations. If cyclin D1 is negative, use Ki67 to exclude Burkitt lymphoma [(BL), usually Ki67 positivity nearly 100 %] and to gauge your suspicion DLBCL (usually Ki67 positivity <80 %) vs. double-hit or double-protein lymphoma (Ki67 usually 80–90 %)

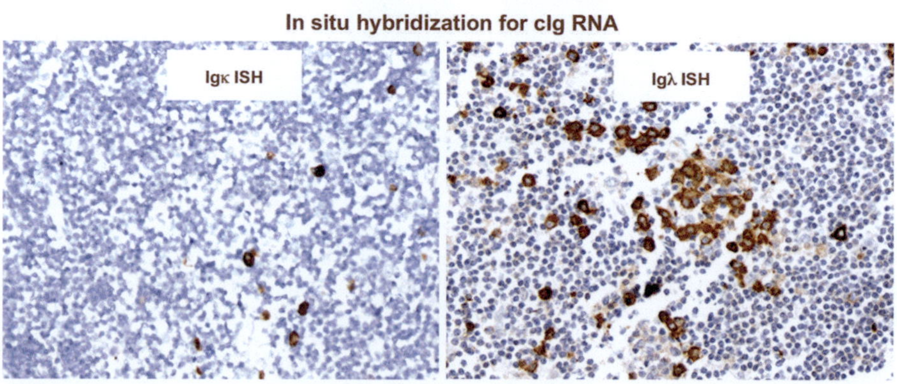

Fig. 10.8 cIg (immunoglobulin) light chain ISH for clonality. Because serum, which contains Ig, is present in the background of all specimens, IHC for Ig protein will always be somewhat problematic, due to staining of the background serum protein. By contrast, Ig RNA only is present within the cytoplasm of plasmacytoid cells. Consequently, Ig light chain ISH is the best, surest way to assess clonality by looking for a subpopulation of lymphoma cells with plasmacytoid differentiation, which will show light chain restriction

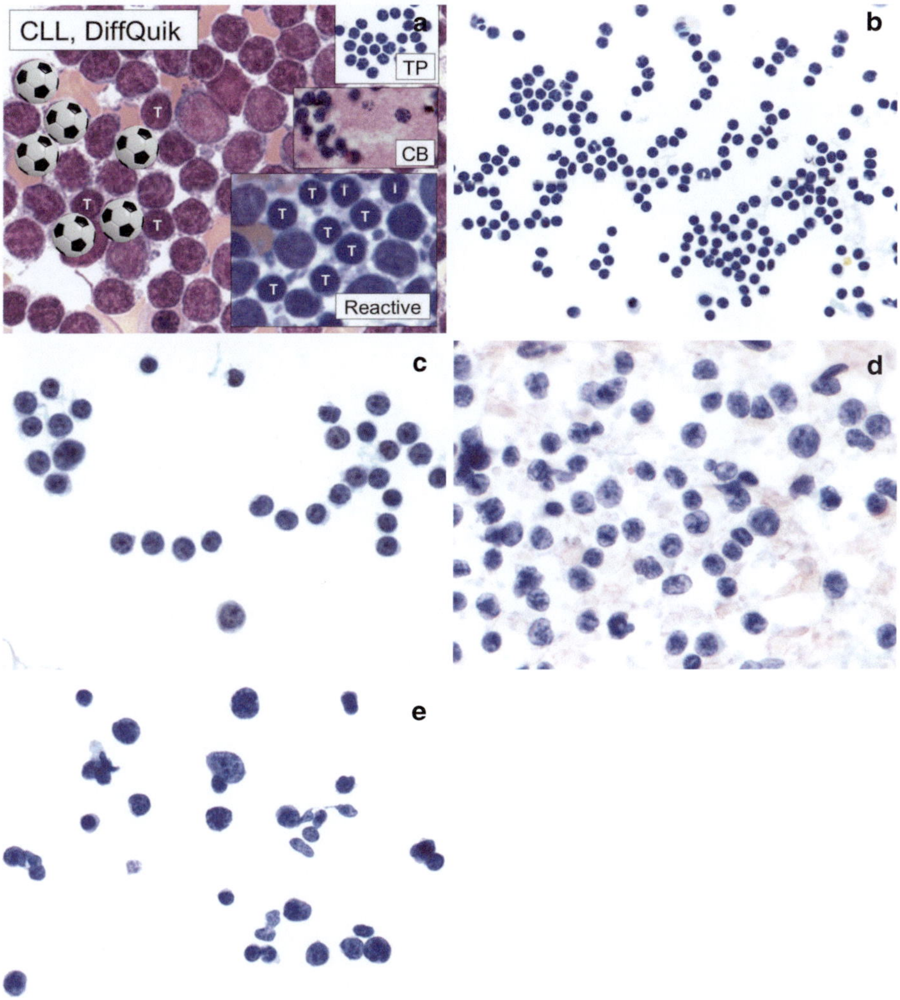

Fig. 10.9 CLL and FL cytology. (**a**) CLL cells resemble reactive small lymphocytes except for the chromatin, which often shows pronounced clumping (likened to soccer balls), as opposed to reactive T cells ("T"), in which the chromatin is typically hyperchromatic and featureless (i.e., chromocenters are less prominent). Compare the DQ-stain features with TP (*top inset*), H&E-stained CB (*middle inset*), and reactive lymphoid cells (*bottom inset*) (DQ-stained DS). (**b** and **c**) CLL cells in TP show morphology as described above. Note "soccer ball-like" chromatin (TP). (**d**) *Low-grade follicular lymphoma* is a challenge to diagnose in cytology. Here the smears show a loose distribution pattern of tumor cells. Lymphoid cells are polymorphous and comprise of two cell types, the centrocytes and centroblasts. The predominant cells are the centrocytes which are small to intermediate-sized lymphocytes (slightly larger than benign lymphocytes) with cleaved nuclei with deep folds in the membrane, occasionally imparting a bilobed appearance. Chromatin is coarsely clumped and nucleoli are not seen. Cytoplasm is scant and basophilic with poorly defined borders. Centroblasts, comprising about 10 % of the cells, are large non-cleaved lymphocytes (twice the size of benign lymphocytes), with round-to-oval nuclei, with occasional indentation. Chromatin is reticular with usually a single or multiple nucleoli. Cytoplasm is scant and dense basophilic or amphophilic. Occasional small mature reactive T lymphocytes are also seen. (**e**) Low-grade follicular lymphoma in TP shows similar features with greater evidence of type and size of neoplastic lymphoid cells. Cytoplasm is basophilic and extremely scant (TP)

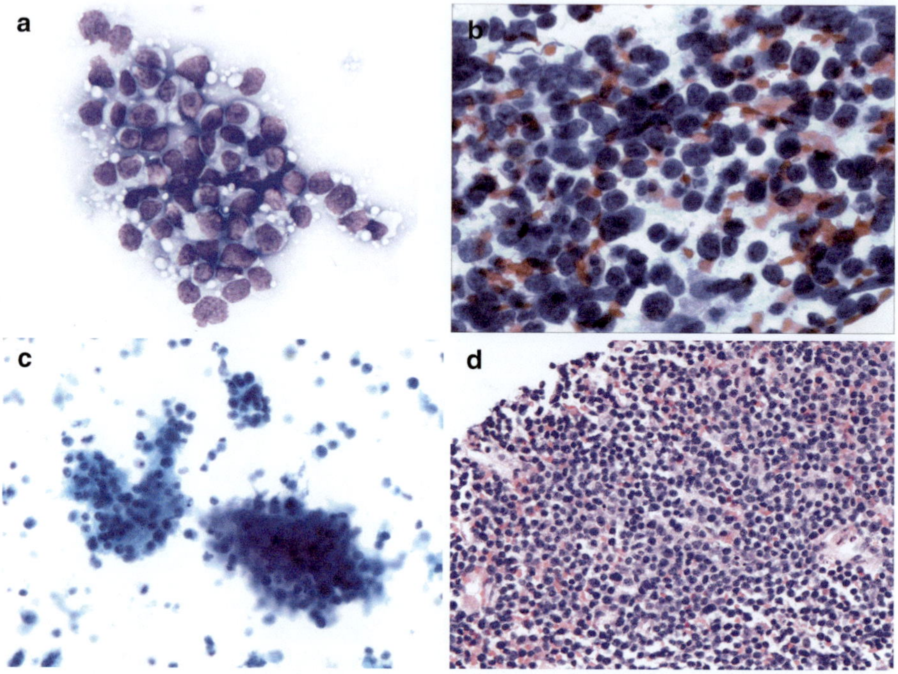

Fig. 10.10 MCL cytology. (**a** and **b**) In MCL the cell size is often difficult to pinpoint (slightly larger than small), but close examination shows that the chromatin is clumped, not truly disbursed. There is no cytologic variability. The infiltrate is monotonous (**a**, DQ-stained DS; **b**, Pap-stained DS). (**c**) In SP, it may only be possible to indicate atypia of lymphoid cells or a suspicion for a low-grade lymphoma based on monotonous population of small lymphoid cells (SP). Correlation with concurrent MFC is recommended. In this case MFC was reported as diagnostic of MCL. (**d**) Histology of MCL shows a diffuse infiltrate of small monotonous lymphoid cells (H&E)

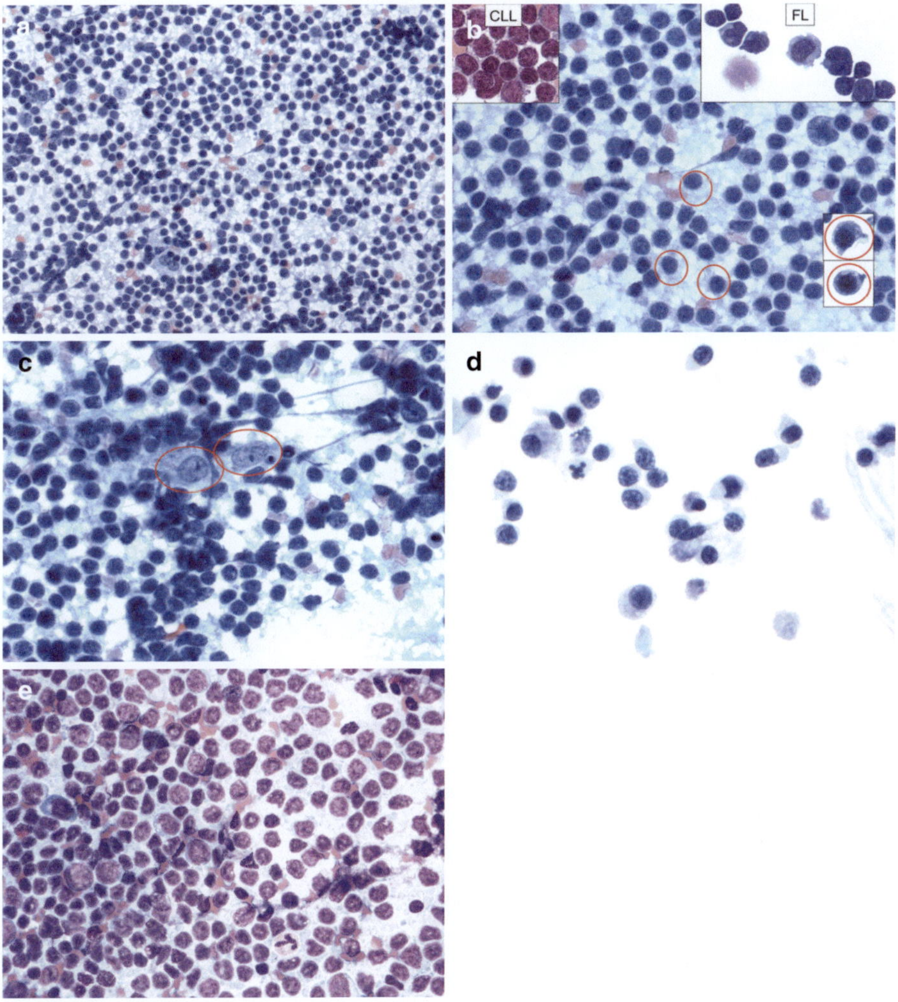

Fig. 10.11 MZL cytology. (**a**) The infiltrate is monotonous and composed of small lymphocytes with round nuclei and clumped chromatin. However, the nuclei are spaced unusually far apart, which suggests the cells have more cytoplasm than normal small lymphocytes DQ-stained DS. (**b**) Higher power confirms the low-power impression of (**a**). Although the cytoplasmic membrane borders are ill-defined, close examination shows that the cells have a monocytoid appearance (*red circles*) due to moderate to abundant cytoplasm; cytoplasmic membranes are more clearly defined. This appearance is distinctly different from CLL (inset, *upper left*) and FL (inset, *upper right*). In FL, there is much greater variability in cell size DQ-stained DS. (**c**) Generally, aggregates of histiocytes (*red circles*) and granulomas indicated a reactive process. However, granulomas can be seen in marginal zone lymphoma. The only other lymphoma that commonly harbors granulomas is Hodgkin lymphoma. (**d**) Note similar morphology in the TP (TP) Pap-stained DS. (**e**) Histology of extranodal MZL (H&E)

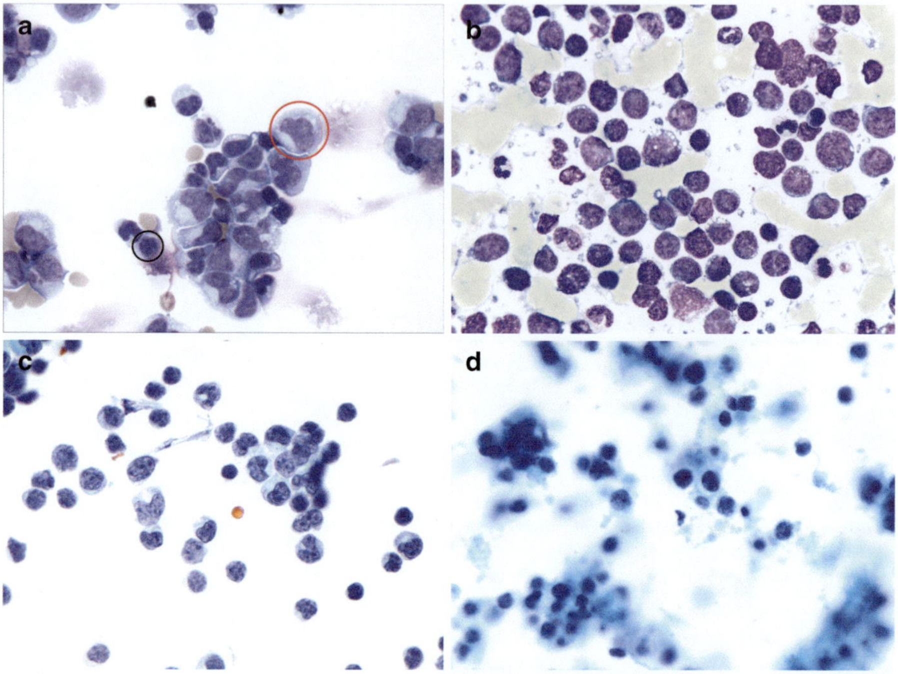

Fig. 10.12 Diffuse large B-cell lymphoma (DLBCL). (**a**) CSF with large cells with a moderate amount of cytoplasm, irregular/lobated nuclei, and disbursed chromatin (*red circles*), strikingly different from reactive small lymphocytes (*black circle*) (DQ-stained cytospin). (**b**) FNA of retroperitoneal lymph node showing DLBCL with necrosis/apoptosis common in aggressive lymphomas, rare in indolent lymphomas (DQ-stained DS). (**c**) DLBCL in an inguinal lymph node FNA. Note the high-grade cytology (TP). Differential diagnosis of DLBCL includes other high-grade lymphomas and poorly differentiated carcinoma. (**d**) A different case of DLBCL processed as SP; note the high-grade cytology and background necrosis (SP)

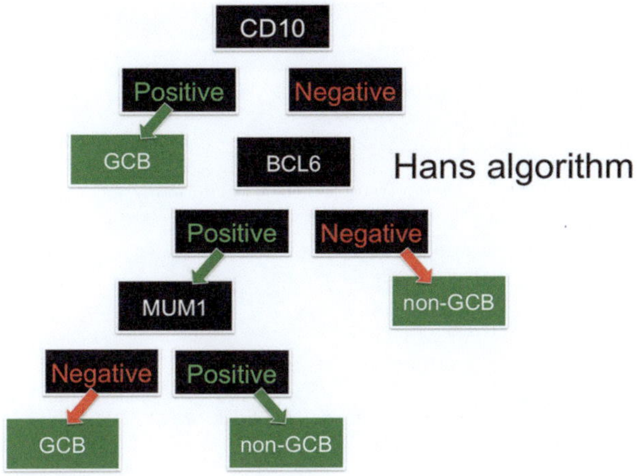

Fig. 10.13 Hans algorithm. Although much has been written about the separation of DLBCL into germinal center (GC) and non-germinal center/activated B-cell (ABC) types, this distinction does not affect therapeutic decisions and has a currently questionable effect on prognosis. However, in many centers the Hans algorithm is still performed. Cytopathologists are not required to have knowledge of this algorithm. It is mainly presented here for awareness

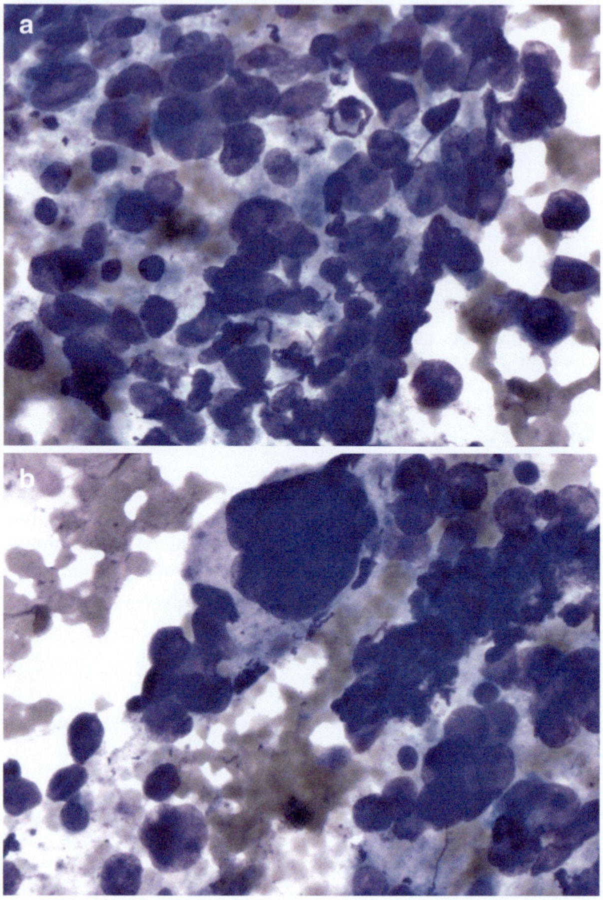

Fig. 10.14 Double-hit lymphoma. (**a** and **b**) this DLBCL was designated as double-hit lymphoma as it showed a Ki67 proliferation index of >80 % and also harbored, both, BCL2 and cMYC gene translocations on FISH (DQ-stained DS)

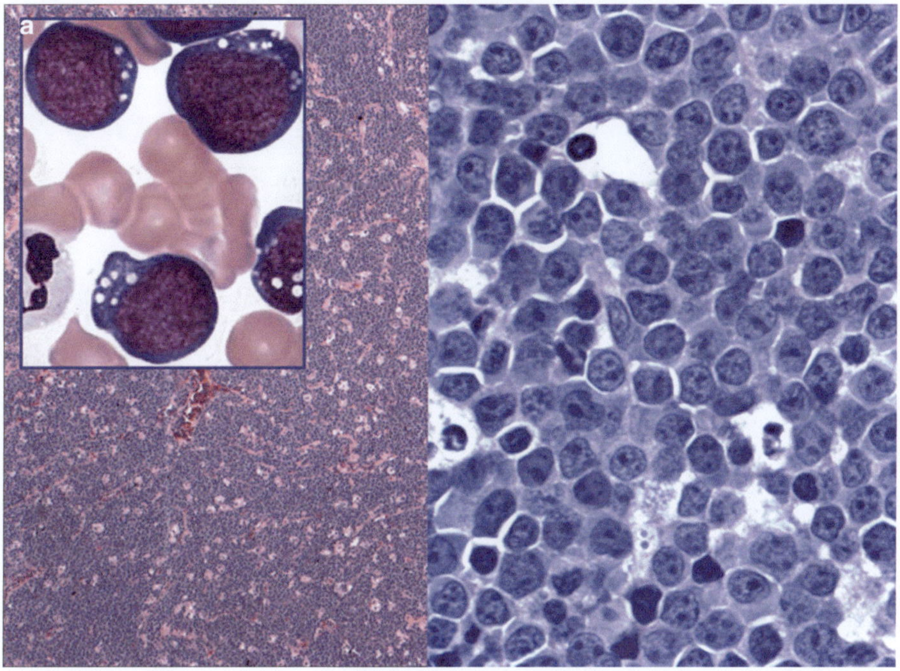

b

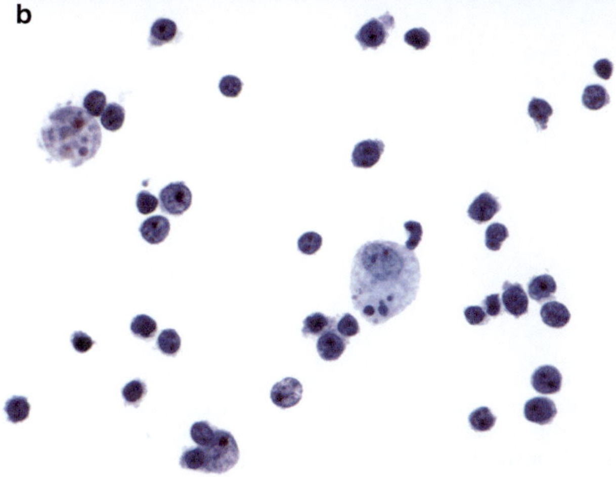

Fig. 10.15 Burkitt lymphoma (BL). (**a**) Main image is a low and high magnification of Burkitt lymphoma (Histology: H&E). The low magnification shows a hematopoietic tumor with numerous tingible body macrophages imparting a "starry-sky" pattern. High magnification shows high-grade lymphoma cells with mostly round nuclei showing indentation. Mitoses and individual cell necrosis (apoptosis) are also seen (H&E). The inset on upper left shows a smear with few monomorphic medium-to-large tumor cells in Burkitt lymphoma. These nuclei are round and uniform with occasional indentations, clumped chromatin and moderate amount of deeply basophilic and vacuolated cytoplasm (DQ-stained DS). Similar cytoplasmic vacuoles can be seen in DLBCL. Some cases of BL may show more nuclear pleomorphism with more prominent nucleoli. (**b**) TP from the same case shows pleomorphic tumor cells with prominent cherry-red nucleoli and interspersed tingible body macrophages (TP)

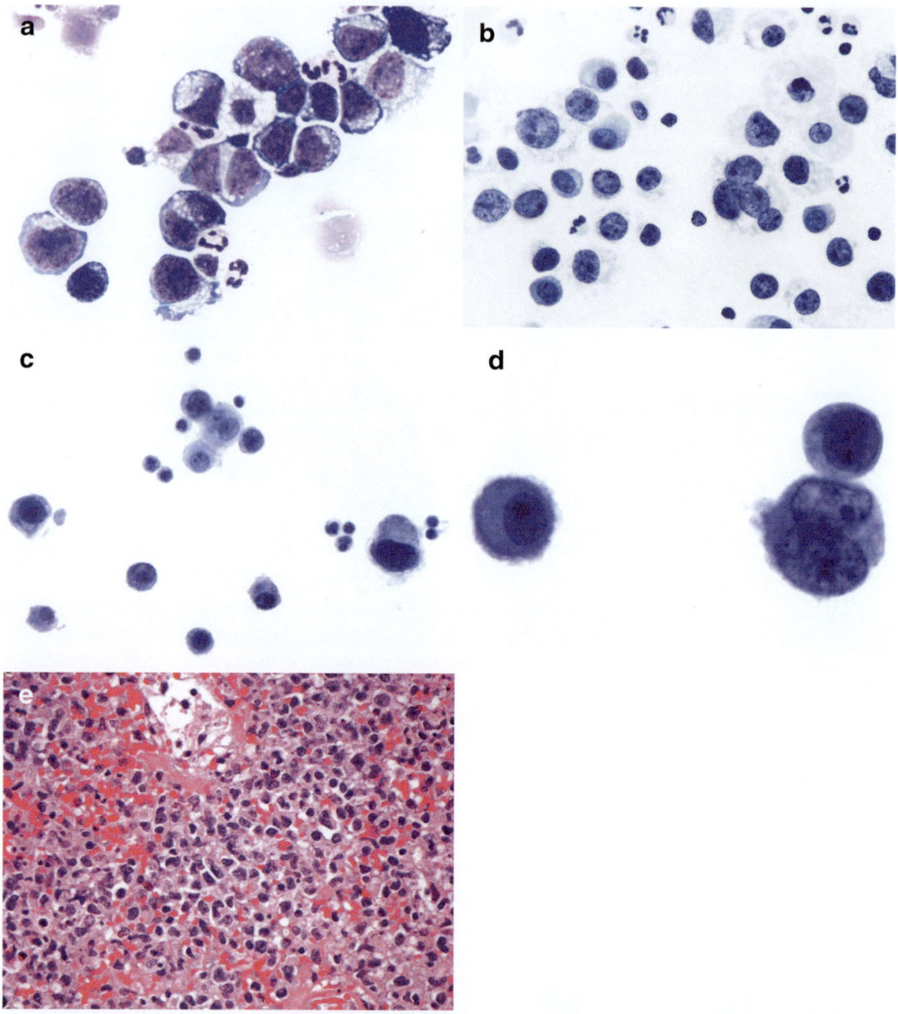

Fig. 10.16 B lymphoblastic leukemia/lymphoma. (**a** and **b**) smears contain small- or intermediate-sized blasts with round-to-convoluted monomorphic nuclei with delicate chromatin (blast-like), distinct nucleoli, and scant-to-moderate weakly basophilic and finely vacuolated cytoplasm (a: DQ-stained DS, b: Pap-stained DS). Lymphoblasts in both B- and T-LBL are positive for TdT. The T-LBL is positive for T-cell antigens CD2, CD3, CD5, CD7, and CD1a, and the B-LBL is positive for B-cell antigens CD19, CD79a, and PAX5. (**c** and **d**) TP shows similar cytological features, with increased variability in size (TP). (**e**) Subsequent excisional biopsy shows diffuse infiltration by medium-to-large neoplastic lymphoid cells admixed with eosinophils (H&E)

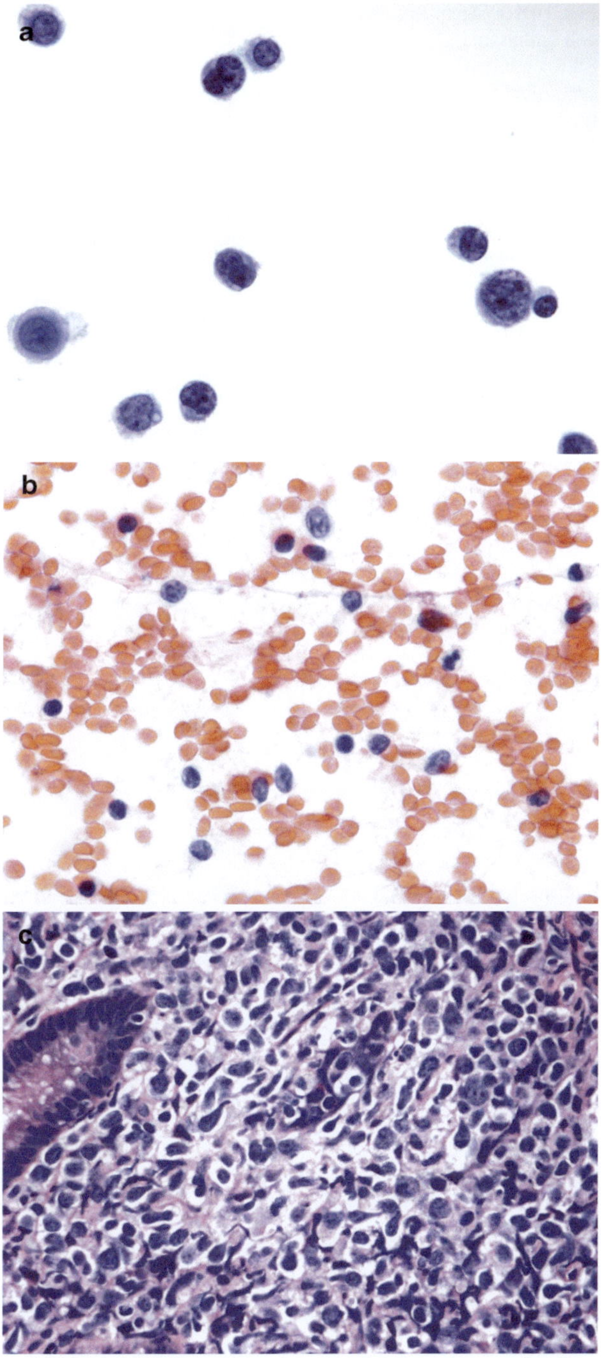

Fig. 10.17 Peripheral T-cell lymphoma. (**a**) Polymorphous populations of small, intermediate, and large neoplastic lymphoid cells are seen. Nuclei are variable in size and exhibit membrane irregularities and nucleoli. Cytoplasm is scant to moderate in amount and pale to basophilic and has occasional vacuolations and well-defined borders. Mitoses and epithelioid histiocytes, eosinophils, and plasma cells may also be seen. (**b**) Similar features are noted in the smear. Nuclear size variability is better appreciated in the smear in comparison to TP, due to the presence of background RBCs (Pap-stained DS). (**c**) Histology of PTCL (H&E)

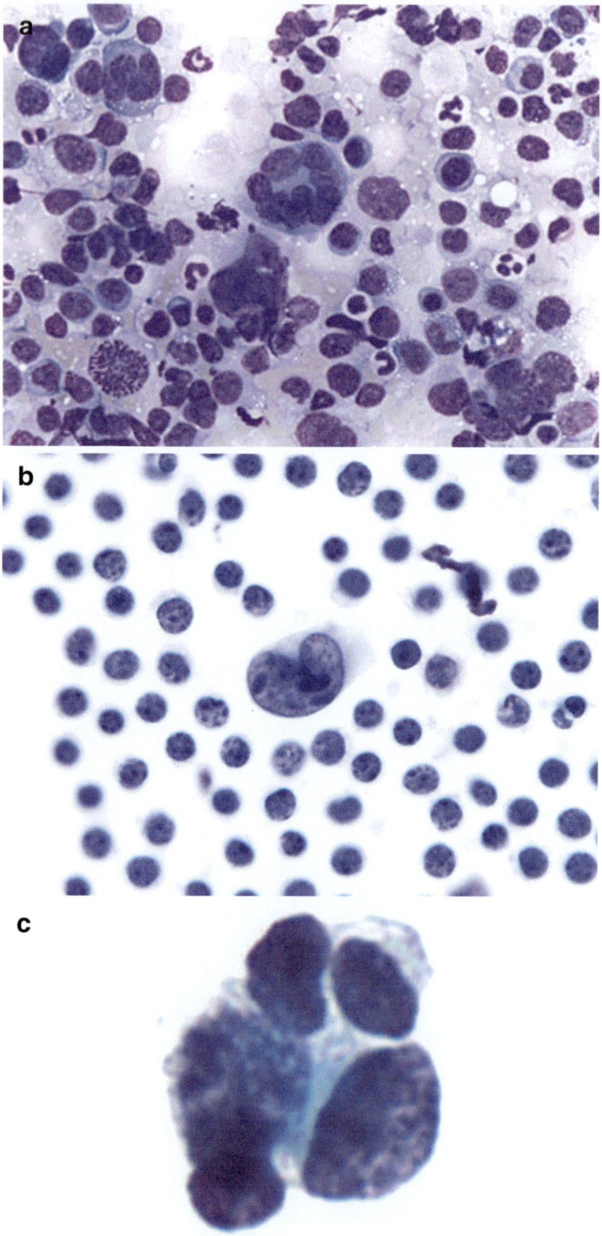

Fig. 10.18 Anaplastic large cell lymphoma (ALCL). (**a**) Shows a broad morphologic spectrum. The center of the field depicts a "doughnut cell" comprising of multi-nucleated giant cells, with nuclei arranged in a wreath-like pattern close to the cytoplasmic border. The upper-left corner depicts the "hallmark" cell of ALCL. The cell is large and pleomorphic with eccentric lobulated (horseshoe-shaped or kidney-shaped) nucleus and abundant basophilic cytoplasm. Multiple Reed-Sternberg-like cells are also present DQ-stained DS. IHC shows positivity for ALK, CD30, EMA, CD45, and one or more T-cell markers including CD2, CD4, and CD5. PCR analyses show T-cell-receptor gene rearrangements, in >80 % of ALK+ ALCL cases. (**b**) "Hallmark" cell on TP. Note the eccentric lobulated (horseshoe-shaped or kidney-shaped) nucleus, with finely granular chromatin, multiple basophilic nucleoli, and faintly basophilic cytoplasm (TP). (**c**) A "doughnut cell" on TP. Note the wreath-like pattern of nuclear arrangement (TP)

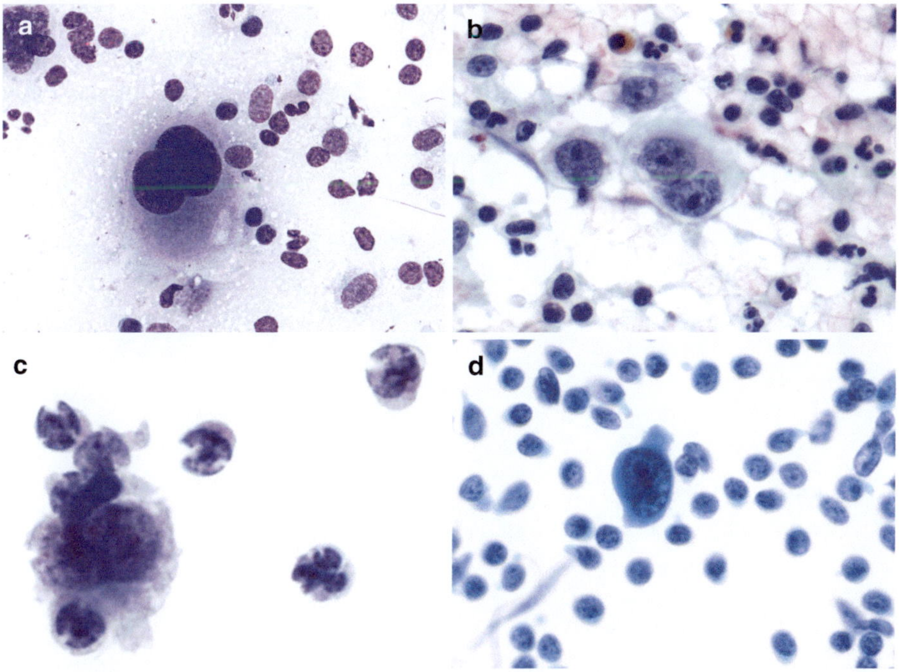

Fig. 10.19 Hodgkin lymphoma (HL). (**a** and **b**) HL in smears show characteristic Reed-Sternberg (R-S) cells. The classic R-S cells are three to four times the size of small lymphocytes and are binucleate or have a bilobed nucleus with prominent eosinophilic nucleoli, giving the cells an "owl-eyed" appearance. Cytoplasm is moderate and basophilic. Small- to medium-sized lymphoid cells, eosinophils, and other inflammatory cells are scattered in the background (a: DQ-stained DS; b: Pap-stained DS). (**c** and **d**) Note HL in TP and SP. R-S cell in the latter have a crisp morphology (**c**, TP; **d**, SP)

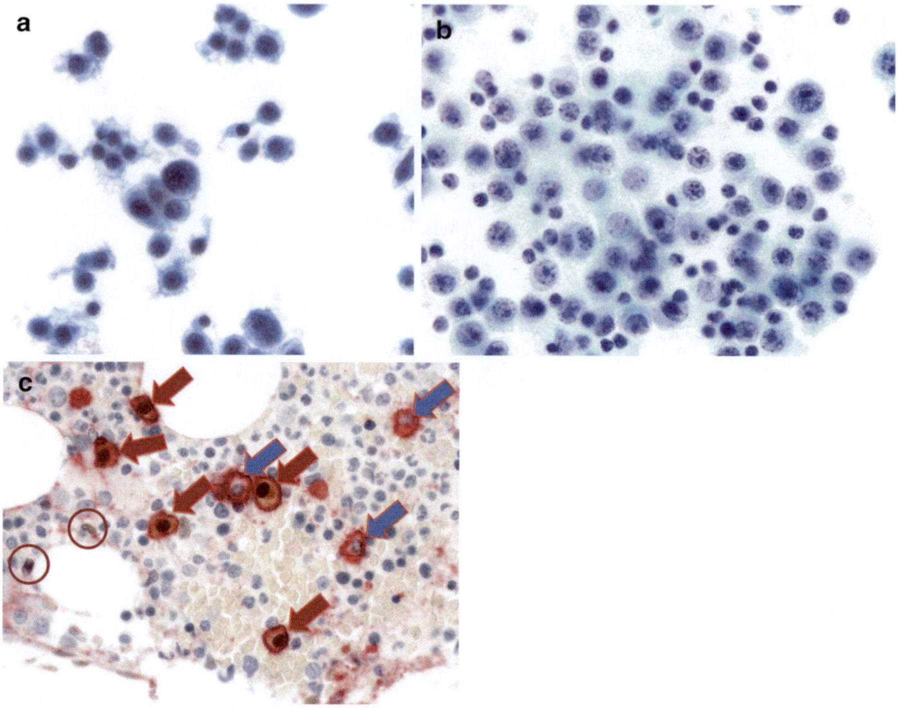

Fig. 10.20 Plasmacytoma. (**a** and **b**) Shows neoplastic plasma cells with varying degrees of maturity. These cells are arranged singly and in sheets and clusters. The mature plasma cells show monomorphic round and eccentric nuclei with condensed "cartwheel" chromatin and indistinct nucleoli. Cytoplasm is abundant and basophilic and shows a perinuclear "hoff" (TP). Binucleated cells are also present. Extracellular eosinophilic material representing amyloid may be seen in some cases, which will show apple-green birefringence with Congo red stain on polarizing light microscopy. (**c**) Cyclin D1 is never expressed by nonneoplastic plasma cells (PC). In this marrow mIHC slide, PCs are identified by red-CD138+ membranes. Coexpression of cyclin D1 (brown nuclei; *brown arrows*) indicates either t(11;14) or hyperdiploidy in a particular PC, both of which are diagnostic of neoplasia. Cyclin D1-negative, nonneoplastic PCs have *blue*, counterstained nuclei (*blue arrows*). By standard, single-stain IHC, it would not be possible to tell with certainty whether the cyclin D1+ nuclei were in PCs or in other cells (e.g., stroma or histiocytes; *brown circles*)

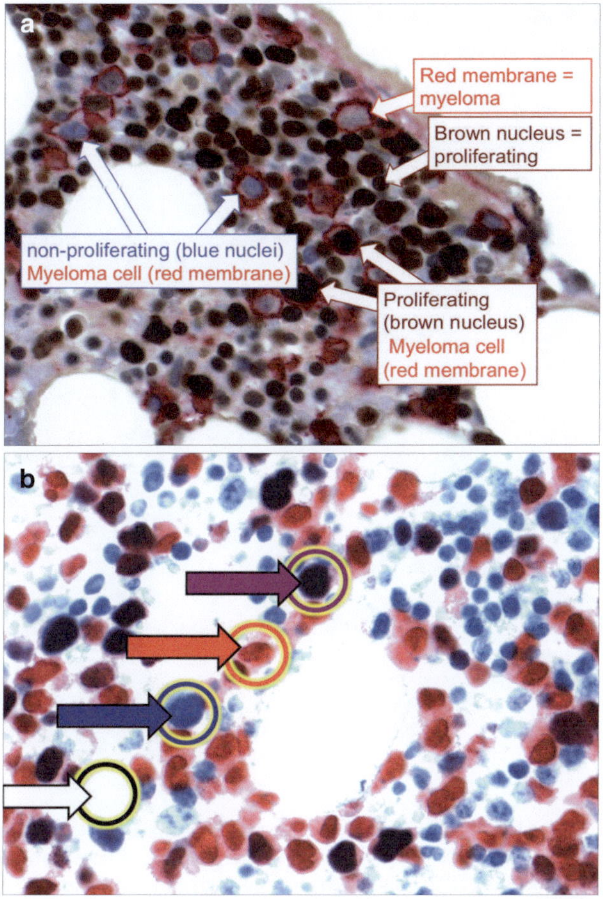

Fig. 10.21 Plasma cell proliferation index (PCPI). (**a**) In this method, the plasma cell proliferation index (PCPI) is performed by counting PCs (*red*, CD138+ membrane) as either proliferating (*brown*, Ki67+ nucleus) or not (*blue*, counterstained nucleus). The index is performed by counting 200 cells and reported as a percentage of proliferating PCs. (**b**) By this method, the plasma cell proliferation index (PCPI) is performed by counting PCs (red, MUM1+ nucleus) as either proliferating (*blue* Ki67+ signal, combined with *red*, MUM1+ signal = purple nucleus; *purple arrow*) or not (*red*, MUM1+/Ki67(−) nucleus; *red arrow*). *Blue*, Ki67+/MUM1(−) cells are background proliferating cells, not included in the analysis. Because a counterstain is not used, double-negative cells are invisible. The index is performed by counting 200 cells and reported as a percentage of proliferating PCs

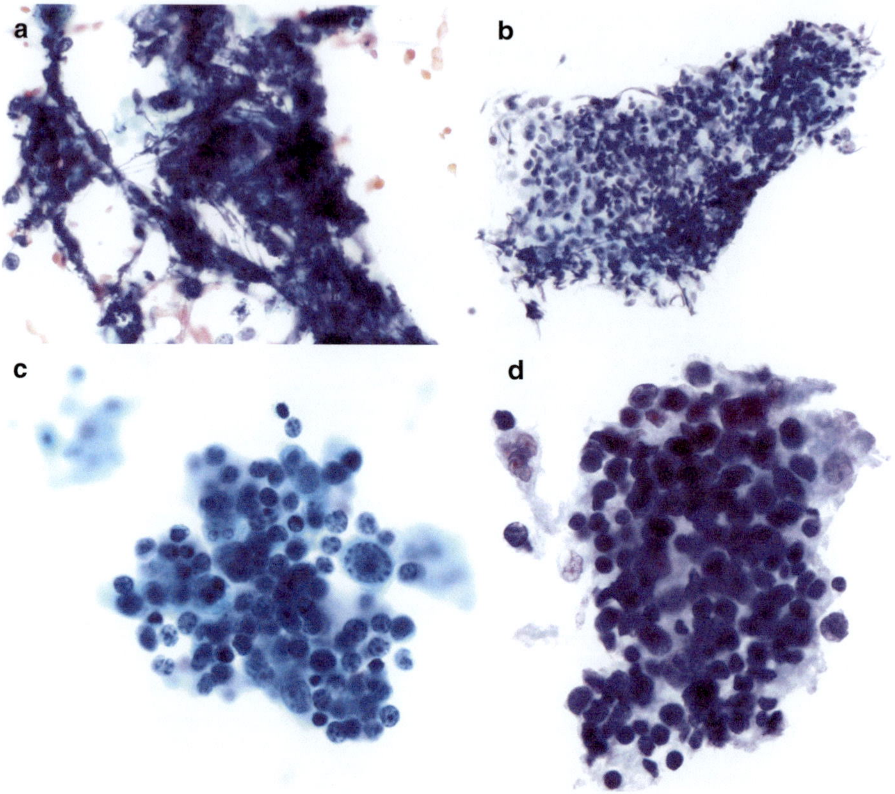

Fig. 10.22 Artifacts of lymph node FNAs. (**a** and **b**) Crush artifact in a low-grade lymphoma can be seen on both smears and LBP. Note that on TP, morphology of some lymphoid cells is still evident (**a**, Pap-stained DS; **b**, TP). (**c** and **d**) Artifactual aggregation or clumping of lymphoid cells may be seen on LBP and lead to diagnostic challenges. In this case, nuclear morphology of lymphoid cells is well preserved on SP despite clumping (**c**, SP; **d**, TP)

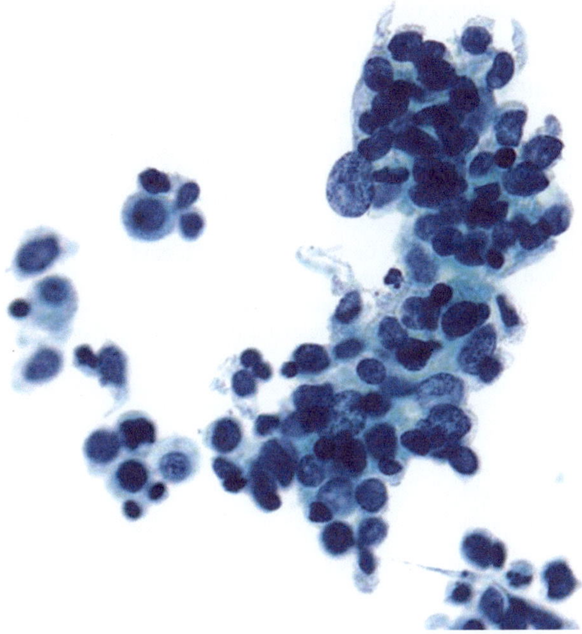

Fig. 10.23 Differential diagnosis of lymphoma in FNA. Small cell carcinoma is the main differential diagnosis. It can be distinguished from lymphoma by nuclear molding, "salt-and-pepper" chromatin, and positive IHC for neuroendocrine markers (TP)

Suggested Reading

1. Swerdlow SH, Campo E, Pileri SA, et al. The 2016 revision of the World Health Organization classification of lymphoid neoplasms. Blood. 2016;127:2375–90.
2. Arber DA, Orazi A, Hasserjian R, et al. The 2016 revision to the World Health Organization classification of myeloid neoplasms and acute leukemia. Blood. 2016;127:2391–405.
3. Smith A, Crouch S, Lax S, et al. Lymphoma incidence, survival and prevalence 2004–2014: sub-type analyses from the UK's Haematological Malignancy Research Network. Br J Cancer. 2015;112:1575–84.
4. Johnson NA, Slack GW, Savage KJ, et al. Concurrent expression of MYC and BCL2 in diffuse large B-Cell Lymphoma treated with Rituximab Plus Cyclophosphamide, Doxorubicin, Vincristine, and Prednisone. J Clin Oncol. 2012;30:3452–9.
5. Rajkumar SV, Dimopoulos MA, Palumbo A, et al. International Myeloma Working Group updated criteria for the diagnosis of multiple myeloma. Lancet Oncol. 2014;15:e538–e48.
6. Swerdlow SHCE, Harris NL, Jaffe ES, Pileri SA, Stein H, Thiele J, Vardiman JW, editors. WHO classification of tumours of Haematopoietic and Lymphoid Tissues Lyon. Lyon: International Agency for Research on Cancer; 2008.
7. Ouansafi I, He B, Fraser C, Nie K, Mathew S, Bhanji R, Hoda R, Arabadjief M, Knowles D, Cerutti A, Orazi A, Tam W. Transformation of follicular lymphoma to plasmablastic lymphoma with c-myc gene rearrangement. Am J Clin Pathol. 2010;134:972–81.
8. Orazi A, Foucar K, Knowles D, Weiss LM. Chapter on role of fine needle aspiration biopsy in the diagnosis and management of Hematolymphoid Neoplasm by Rana Hoda, in Knowles' neoplastic hematology. 3rd ed. Philadelphia Wolters Kluwer; 2013. p. 293–321.
9. Barrena S, Almeida J, Del Carmen G-MM, López A, Rasillo A, Sayagués JM, Rivas RA, Gutiérrez ML, Ciudad J, Flores T, Balanzategui A, Caballero MD, Orfao A. Flow cytometry

immunophenotyping of fine-needle aspiration specimens: utility in the diagnosis and classification of non-Hodgkin lymphomas. Histopathology. 2011;58:906–18.
10. Ali AE, Morgen EK, Geddie WR, et al. Classifying B-cell non-Hodgkin lymphoma by using MIB-1 proliferative index in fine-needle aspirates. Cancer Cytopathol. 2010;118:166–72.
11. Pambuccian SE, Bardales RH. Lymph node cytopathology; essentials in cytopathology. New York: Springer; 2011.
12. Savage EC, Vanderheyden AD, Bell AM, et al. Independent diagnostic accuracy of flow cytometry obtained from fine-needle aspirates: a 10-year experience with 451 cases. Am J Clin Pathol. 2011;135:304–9.
13. Zhang S, Abreo F, Lowery-Nordberg M, et al. The role of fluorescence in situ hybridization and polymerase chain reaction in the diagnosis and classification of lymphoproliferative disorders on fine-needle aspiration. Cancer Cytopathol. 2010;118:105–12.
14. Jiménez-Heffernan JA, Vicandi B, López-Ferrer P, et al. Value of fine needle aspiration cytology in the initial diagnosis of Hodgkin's disease. Analysis of 188 cases with an emphasis on diagnostic pitfalls. ActaCytol. 2001;45:300–6.

Pancreas and Liver Fine Needle Aspiration

11

Introduction

The gastrointestinal (GI) tract is comprised of an upper and lower tract and associated accessory organs. Cytology of the GI tract is performed in order to detect neoplastic and non-neoplastic lesions and often includes concurrent procurement of needle care biopsy (NCB). Sites amenable to cytological sampling include the esophagus, stomach, duodenum, pancreatobiliary and liver from the upper GI tract, rectum and anus from the lower GI tract, and regional lymph nodes. See Chap. 4, "Gastrointestinal Tract Exfoliative Cytology," for introductory details, PSC reporting guidelines, and ancillary tests for GI specimens.

Indication, Collection, and Laboratory Processing of Cytological Samples

The examination of GI cytology is an efficient and cost-effective method for diagnosing GI tract lesions. It allows for rapid interpretation of FNA specimens using rapid on-site evaluation (ROSE); this also allows for proper specimen triage for various ancillary tests. An accurate diagnosis relies on procuring an adequate sample, optimal specimen preparation in order to enhance cytomorphology, and expertise in the interpretation of these challenging entities.

Endoscopic Ultrasound-Guided FNA (EUS-FNA)

Endoscopic ultrasound-guided fine needle aspiration (EUS-FNA) is a well-established, reliable, accurate, and safe diagnostic technique for assessing lesions of the GI tract and adjacent organs. EUS-FNA can also be used to sample mediastinal, splenic, and adrenal gland lesions. The sensitivity and specificity of EUS-FNA are at least 80 and 100 %, respectively.

R.S. Hoda et al., *Diagnostic Liquid-Based Cytology*,
DOI 10.1007/978-3-662-53905-7_11, © Springer-Verlag GmbH Germany 2017

EUS-FNAs are usually performed for masses in the esophagus, stomach, esophageal/gastric submucosal lesions, pancreatobiliary tract, and regional lymph nodes including periesophageal/gastric, peri-pancreatobiliary, and abdominal lymph nodes. It is the procedure of choice for establishing a pancreatic malignancy and for sampling pancreatic cysts.

EUS-FNA of pancreatic masses is performed with linear endosonographic instruments. The aspiration needles are usually 22- or 25-gauge and are selected based on the vascularity of the lesion or viscosity of cyst fluid. EUS-FNA may be performed in conjunction with NCB (Tru-Cut 22-gauge Shark Core needles) for mucosal, submucosal, fibrotic, and stromal-rich lesions. ROSE of cytology is usually performed for FNAs of both solid masses and cystic lesions, as it reduces rate of false-negative FNAs. After the cyst fluid is analyzed cytologically, cyst fluid is triaged for biochemical testing [carcinoembryonic antigen (CEA) and amylase] and/or molecular testing for mutation analysis. Additional passes are performed if a residual solid component (mural nodule) is identified.

EUS is better than CT or MRI in determining vascular invasion and detecting small tumors and lymph node metastases.

Laboratory Processing of GI Specimens

For FNA, the specimen is processed as conventional smears, LBP, and/or a cell block. An accurate cytological diagnosis requires an adequate and well-prepared sample, as well as expertise in the interpretation of these preparations. See Chap. 1 for details of LBP preparations.

Advantages of FNA Specimens for GI Tract over NCB

Ability to assess solid pancreatic lesions; subepithelial, submucosal, and mural nodules; and regional lymph nodes via EUS-FNA, assessment of cystic lesions, collection of cyst fluid for chemical and molecular analyses, and shorter turnaround time.

EUS-FNA of Submucosal Gastric and Esophageal Lesions

EUS-FNA is an effective method for tissue diagnosis of gastrointestinal submucosal tumors (SMTs). The standard endoscopic NCB is usually nondiagnostic because the overlying mucosa is usually thick and prevents procurement of tissue from the lesion. The SMTs include gastrointestinal stromal tumors (GISTs). The diagnosis of GIST is based on cytomorphology and immunohistochemical analysis. The latter is crucial for differentiating these lesions from benign smooth muscle lesions such as a leiomyoma or schwannoma. Therefore, it is necessary to obtain adequate tissue for the immunohistochemical analysis in the case of GI SMTs. Some surprises may be encountered such as esophageal duplication cyst.

EUS-FNA of Regional Lymph Nodes

EUS-FNA is an important modality for sampling regional deep-seated lymph nodes, including mediastinal and abdominal, for diagnosis of infections, metastatic malignancy, or primary hematopoietic tumors. Adequate specimen should be procured for ancillary tests including immunohistochemistry, flow cytometry, and molecular analyses, such as next-generation sequencing (when pertinent), to increase diagnostic yield. Lymphoid lesions may be successfully subclassified on EUS-FNA, particularly, if the lymphoma is recurrent. Cytopathologists should work in collaboration with hematopathologists for work-up of lymphoid lesions.

Pancreatic Lesions

The differential diagnosis of solid and cystic pancreatic lesions include:

- Solid
  - Chronic pancreatitis
  - Pancreatic adenocarcinoma
  - Neuroendocrine tumors
  - Solid pseudopapillary tumor
  - Acinar cell carcinoma
- Cystic, neoplastic
  - Mucinous
    - Intraductal papillary mucinous neoplasms (IPMNs)
    - Mucinous cystic neoplasm (MCN)
  - Non-mucinous
    Serous cyst
- Cystic, non-neoplastic
  - Pseudocyst

Solid Pancreatic Lesions

Any of the above-noted solid pancreatic lesions may be encountered on EUS-FNA. The clinician performing the procedure may have a suspicion for a particular entity based on the clinical history and presentation, preceding CT and MRI findings and appearance on EUS. Fortunately, each entity has specific cytological features and distinct immunoprofile and occasionally a distinct location. Common diagnostic pitfalls include chronic pancreatitis that may mimic well-differentiated pancreatic adenocarcinoma and vice versa and benign acinar cells that may mimic neuroendocrine (NE) tumor and vice versa.

Pancreatic Cysts

Cystic pancreatic lesions are detected by imaging techniques and are sampled by EUS-FNA, when clinically and radiologically indicated. Clinically, it is important to determine if the pancreatic cyst is mucinous or non-mucinous and benign or malignant. Inflammatory pseudocysts can be managed medically in most patients, and serous cysts which are benign neoplasms are not surgically resected unless the patient is symptomatic or the cyst is large (>4 cm). Neoplastic mucinous cyst can be IPMN or MCN. Although both are mucinous epithelium-lined cystic tumors, they differ in age, gender, location in the pancreas, relationship to pancreatic ducts, and management. All main pancreatic duct IPMNs are resected. Management guidelines for suspected branch duct-type IMPNs are based on clinical and imaging findings. Small cysts (<3 cm) without high-risk clinical or imaging findings are conservatively managed. Patients with high-risk imaging features or suspicious/malignant cytology are surgically managed. Patients with worrisome imaging features will be evaluated by EUS-FNA for cytological assessment. All MCNs are resected for assessment of high-grade dysplasia. Aspiration of pancreatic cysts is also valuable for obtaining material for cytological, biochemical, and molecular analyses.

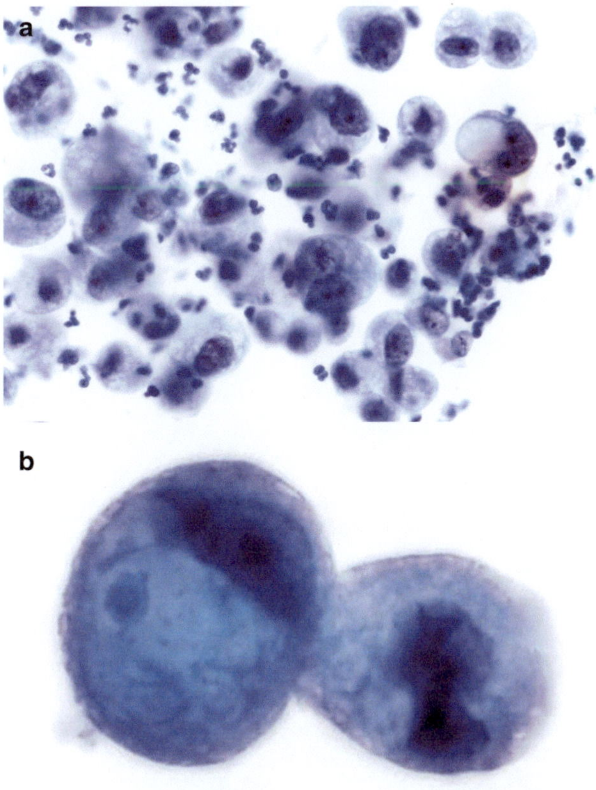

Fig. 11.1 Diffuse signet ring-type gastric adenocarcinoma. (**a**) EUS-FNA yielded this material from a gastric lesion. This pattern is classic for signet ring gastric cancer, in which some cells take on the appearance of a signet ring (see the cell at 2 o'clock) – the "hole" of the ring is a mucin vacuole, which appears mostly white and the "ring" is the thin rim of cytoplasm. The malignant nucleus is large, is eccentrically placed, and appears to bulge out of the cell, completing the picture of a signet ring. These carcinomas often diffusely involve the stomach as single cells, and thus FNA often produces numerous discohesive single cells rather than tissue fragments. Note the nuclear size differences, the coarse chromatin, and the irregular nuclear borders seen in these cells. These cells may resemble histiocytes, and a high degree of suspicion may be required for their detection. In this direct smear (DS), inflammatory cells partially obscure cell detail (Pap-stained DS). (**b**) Isolated signet ring cells with the crescentic-shaped nuclei compressed toward the periphery by large intracytoplasmic mucinous vacuole, characteristic of the tumor. Mucinous cytoplasmic condensation is evident in the cell on the left. Background is clean (TP). If cell block material is present, keratin markers or special stain for mucicarmine can differentiate the cells from histiocytes, which express CD68 or KP-1 [1]

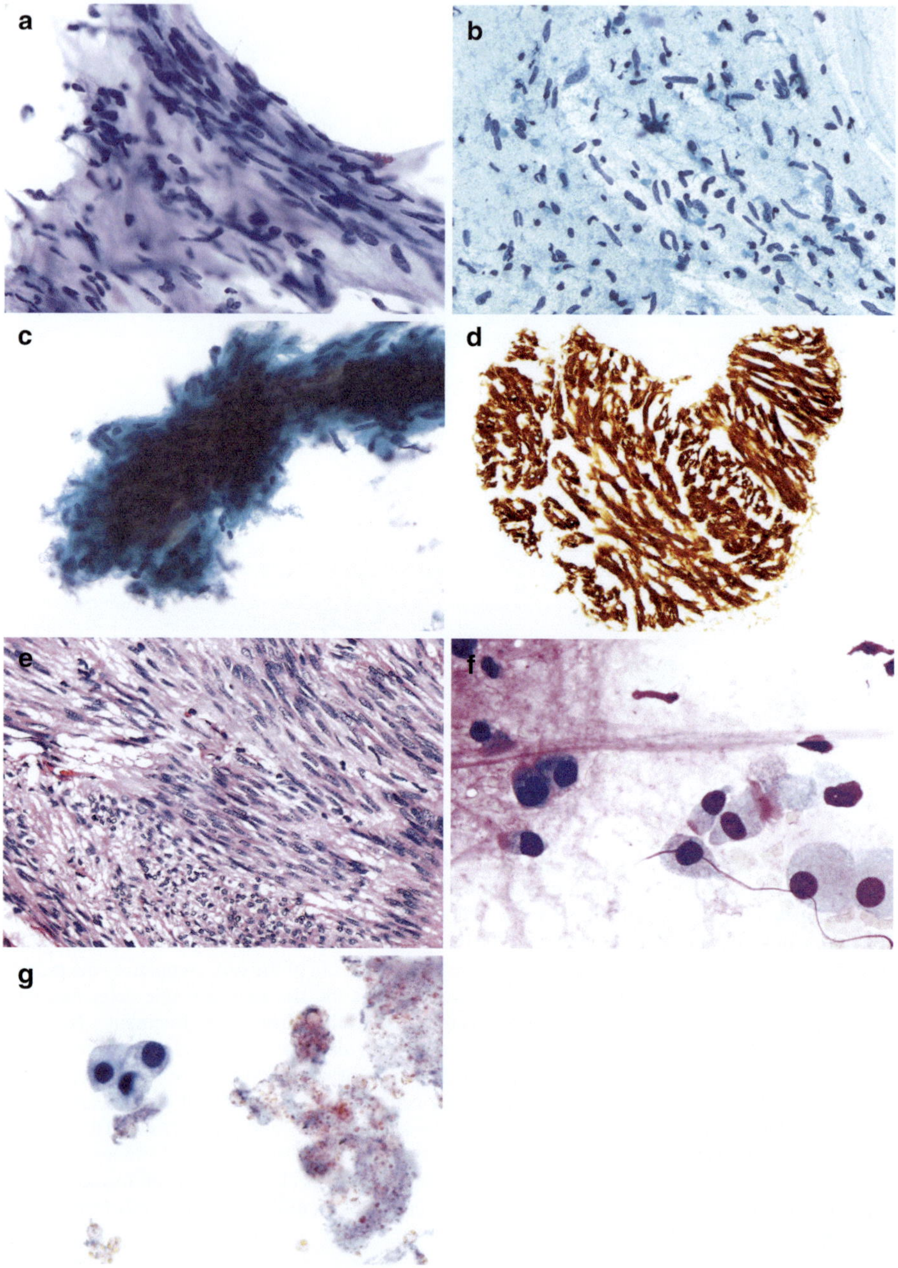
a
b
c
d
e
f
g

Fig. 11.2 EUS-FNA of gastrointestinal stromal tumor (GIST). (**a**) This aspirate shows numerous uniform spindle cells with monotonous, slender, well-aliened nuclei with pale chromatin, no nucleoli, and delicate wispy basophilic cytoplasm. The cytoplasmic borders are indistinct, giving a syncytial appearance. The fragment is too cellular to be a fragment of normal connective tissue, which is sometimes seen when a lesion is not properly sampled during the procedure. Note the myxoid stromal quality. Mitoses were not evident (TP). (**b**) GIST on a conventional smear shows similar nuclear morphology. The myxoid stroma is more diffuse (Pap-stained DS). (**c**) Same case of GIST as above processed as SP specimen. Main differences are that the stroma appears denser and cells are more compact than TP (SP). (**d**) Positive immunostain for c-kit (a cell-surface transmembrane tyrosine kinase also known as CD117). Immunostain for DOG1 was also positive (CB). (**e**) The resection shows fascicles of uniform, bland spindle cells with pale, eosinophilic cytoplasm and myxoid stroma (H&E). Radiologically, this lesion was submucosal, distal, well marginated, heterogeneous, and enhancing. GISTs are the most common mesenchymal neoplasms of the gastrointestinal tract and should be differentiated from other mesenchymal tumors. They harbor specific activating mutations in the *KIT* or platelet-derived growth factor receptor-α (*PDGFRA*) which makes them responsive to pharmacologic inhibitors, such as imatinib mesylate, and sunitinib malate. Overall, it is difficult to completely distinguish GIST from other spindle-shaped neoplasms in this region, and immunohistochemical studies are often critical for a definitive diagnosis. Approximately 90 % of GISTs are positive for c-kit and 95 % are positive for DOG-1 [2]. S-100 positivity is suggestive of a nerve sheath tumor, such as schwannoma, whereas smooth muscle actin (SMA) reactivity is seen in smooth muscle neoplasms, such as leiomyoma. Clinical suspicion for GIST may arise from imaging studies showing a submucosal location of the tumor. (**f** and **g**) *Gastric duplication cyst*; a second lesion was identified as a well-defined, 3-cm hypoechoic lesion with a heterogeneous internal structure, along the greater curvature of the stomach/peripancreatic by a previously performed computed tomography (CT) scan. It was suspected to be a lymph node suspicious for non-Hodgkin lymphoma. The cystic nature of the lesion was not diagnosed due to the solid-tissue appearance of the lesion, presumably, because of the mucinous nature of the fluid (DQ). The images demonstrate thick mucin with embedded ciliated columnar epithelial cells. Mucin is more diffuse and stains magenta on DQ and clumped and pale basophilic on TP (**f**, DQ-stained DS; **g**, TP). Histologically, the cysts were lined with pseudostratified ciliated columnar as seen on subsequent resection

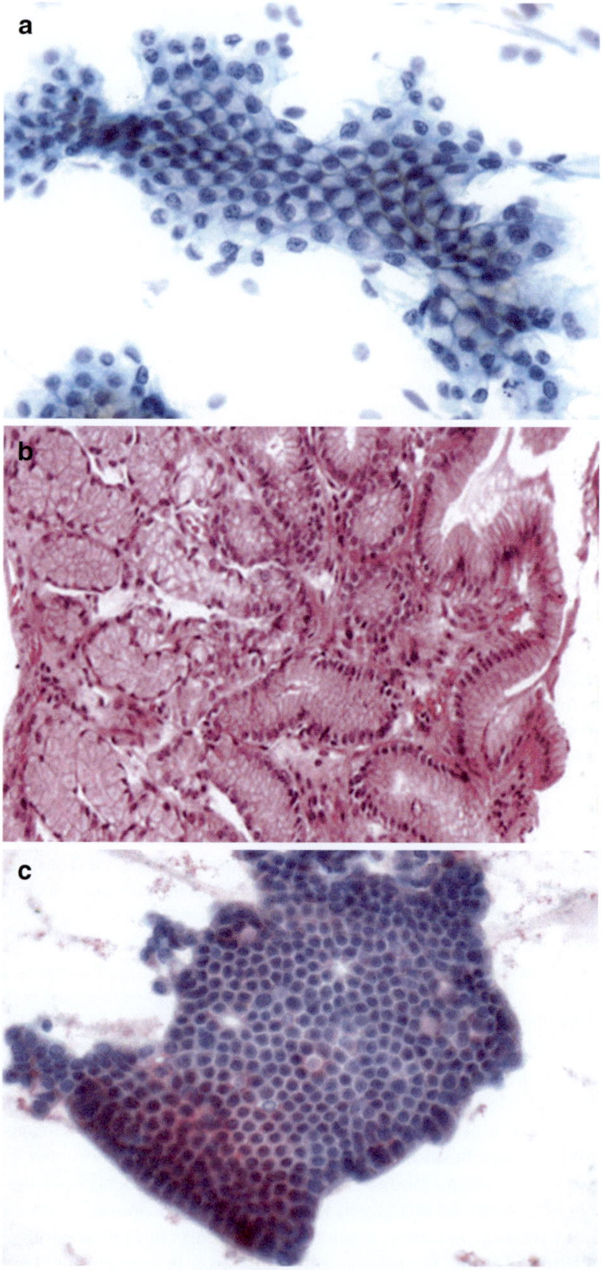
a
b
c

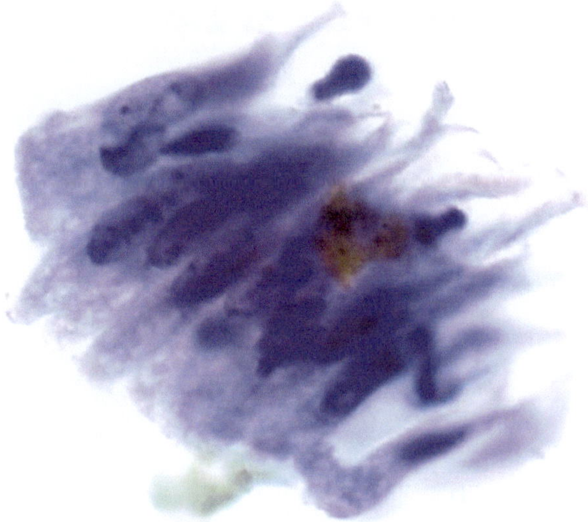

Fig. 11.4 Adenoma of duodenal ampulla. Strips of columnar glandular cells in which the cells maintain nuclear polarity. The nuclei are elongated but remain opposite of the luminal border of this fragment (TP). Villous adenoma of the ampulla is rare, and the distinguishing cytomorphological characteristics have been reported, in DS, to be slender, elongate columnar cells with long, thin, basally oriented nuclei, nucleoli, and fine, granular chromatin [3]. LBP show similar morphology

Fig. 11.3 Benign gastric and duodenal mucosa. (**a**) Benign gastric foveolar cells showing a honeycomb arrangement with an apical mucin cup (TP). (**b**) Benign gastric mucosa with branching mucus glands with vacuolated cytoplasm (H&E). (**c**) Benign duodenal mucosa appears as flat, two-dimensional sheet with an organized honeycomb configuration. Cells are polarized and uniform with low nucleus-to-cytoplasmic ratios. The nuclei are monotonous with bland chromatin. Most importantly, duodenal mucosa shows goblet cells, which appear as evenly distributed "eosinophilic holes" in the fragments which are not seen in pancreatic mucinous neoplasms (TP). EUS-FNA of the pancreas must transverse the normal GI tract mucosa and may be seen as contaminants in pancreatic FNA specimens. EUS-FNA for pancreatic head and body/tail lesions are performed via the duodenum and the stomach, respectively. Because the GI contaminants are also mucin-secreting epithelium, one must carefully distinguish them from cystic mucinous tumors

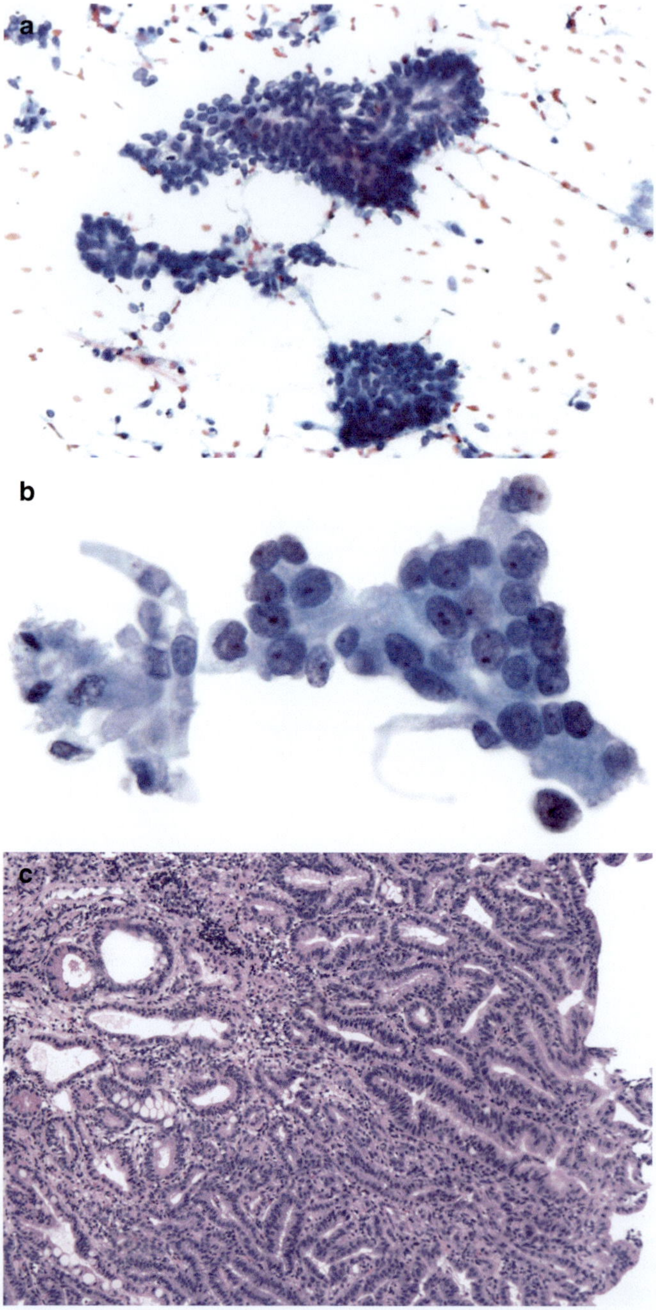

Fig. 11.5 Duodenal ampullary adenocarcinoma. (**a**) Highly cellular smear with both fragments and discohesive single cells that suggest a neoplastic lesion. The fragments are arranged in a glandular formation and do not contain goblet cells (Pap-stained DS). (**b**) A TP preparation of the same case shows a glandular formation with malignant nuclei. Elongated cells as seen in the adenoma case in Fig. 11.4 are also present. Compared to the DS, single cells are rare (TP). (**c**) Ampullary adenocarcinoma showing an infiltrative pattern of highly irregular glands with dark nuclei (H&E)

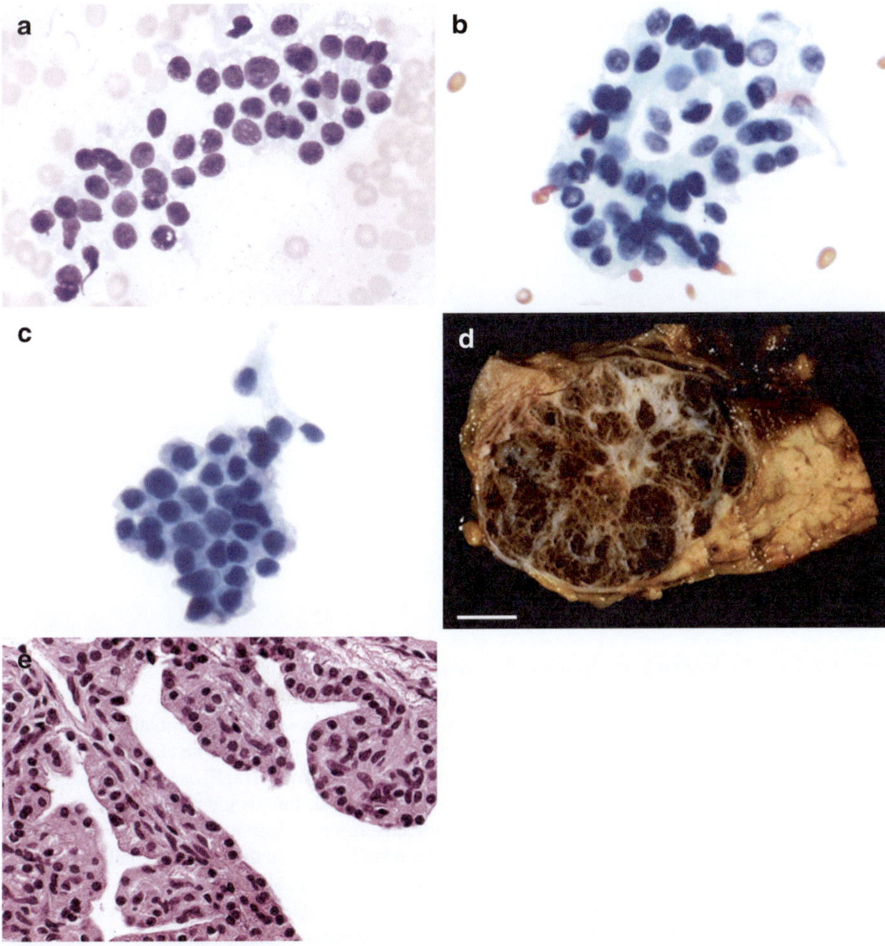

Fig. 11.6 EUS-FNA of a serous cystadenoma. (**a** and **b**) Cells with bland-appearing nuclei. Cytoplasm which appears finely vacuolated on DQ stain and pale blue on Pap stain. Nuclei are uniform and bland with no nucleoli and are not much larger than RBCs. These cells could be mistaken for pancreatic acinar cells or islet cells (**a**, DQ-stained DS; **b**, Pap-stained DS). (**c**) Cells in TP appear similar to those seen in (**a**), except they are more compact (TP). (**d**) Gross of serous cystadenoma displaying a large well-defined microcystic tumor with a central stellate scar. (**e**) Histology of serous cystadenoma shows small cystic spaces lined by small cuboidal cells with clear glycogen-containing cytoplasm. EUS-FNA has very poor performance for detecting serous cystadenomas, in part because the cyst lining is not always sampled, and when sampled the cyst lining is typically scant and may be mistaken for other bland-appearing elements in EUS-FNA of the pancreas. One study has reported that less than 20 % of cases contained the characteristic serous lining cells [4]

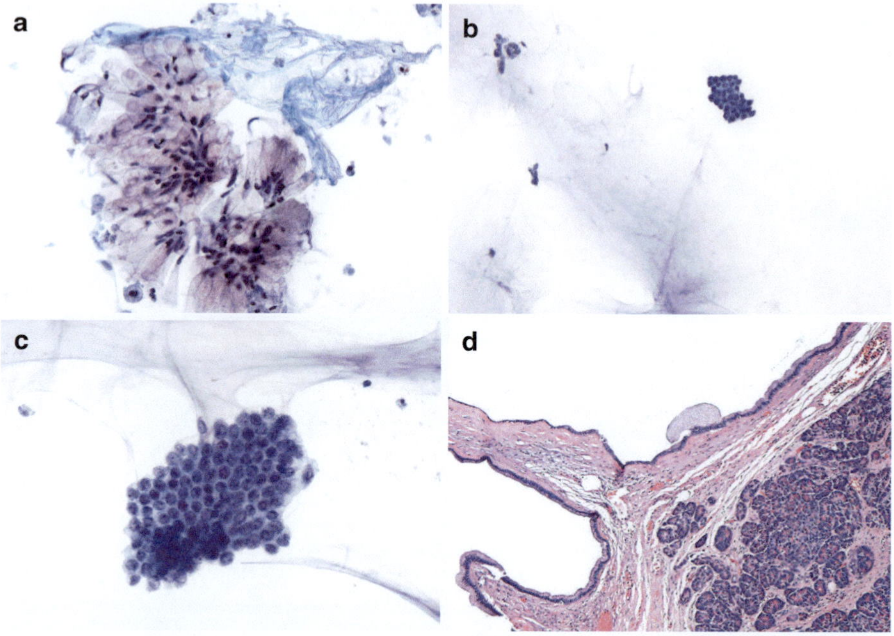

Fig. 11.7 EUS-FNA of an intraductal papillary mucinous neoplasm (IPMN). (**a**) Mucinous cells with thick extracellular mucin. Note benign nature of cells (Pap-stained DS). (**b**) This field demonstrates one of the most important features of an IPMN: "clean" mucin (as opposed to contaminating mucin from the gastrointestinal tract, which often contains degenerated cells and bacteria) with clusters of epithelial cells floating within or close to it. The presence of abundant "clean" mucin on FNA is at least suspicious for a mucin-producing cyst, and in the correct radiologic context, this would support a diagnosis of IPMN, even if the glandular lining component was under-sampled. Clean mucin can also be found in mucinous cystic neoplasm (MCN), though these lesions usually occur in the pancreatic tail; IPMNs are more commonly found near the pancreatic head (TP). (**c**) Gastric foveolar-type cells of IPMN with low-grade dysplasia (LGD). The cells have small nucleoli, mild anisonucleosis, mild nuclear border irregularities, scattered intranuclear grooves, and rare inclusions. Note the "clean" mucin which surrounds these cells in the background (TP). One pitfall of LBP with cystic mucinous lesions is that mucin may not be present in the background, leaving interpretation to rely on the presence of mucinous lining alone [5]. (**d**) Most of the mucinous contents of this IPMN have been removed by processing, though some small amount of mucin remains attached to the mucinous epithelial lining. This appears to be a large cyst, but the cystic lining is the neoplastic element. Normal pancreatic elements underlie the neoplastic cyst. While connection to the pancreatic duct can only be demonstrated grossly, this lesion lacks the ovarian stroma that typically surrounds a mucinous cystic neoplasm (MCN) (H&E)

Fig. 11.8 EUS-FNA of an IPMN with high-grade dysplasia. (**a**) Cells are variable in size and appear in small sheets, in groups, and as single cells. The nuclei are enlarged, round, and hyperchromatic with prominent nucleoli and high nucleus-to-cytoplasmic ratio. Cytoplasm is scant and shows a single mucinous vacuole. Mitoses and necrosis were present. Some cells appear in small bud-like arrangement (TP). (**b**) Another characteristic feature comprises of cluster of tumor cells with overcrowding and overlapping. Nuclei appear vesicular with small cherry-red nucleoli and intranuclear grooves. Mucin or mucin-containing cytoplasm is not evident (TP). (**c**) The corresponding resection specimen demonstrating IPMN with high-grade dysplasia (H&E). Grading of dysplasia in IPMN on FNA is difficult, in part because it depends on the sampling of the cyst lining and secondly because of the morphological overlap between lesions of different grade as well as invasive adenocarcinoma. Sigel et al. [6] demonstrated that a three-tier system resulted in improved agreement among observers. This system included a nondiagnostic category, a category which combined low-grade dysplasia and atypia, and a third category which combined high-grade dysplasia, suspicious for adenocarcinoma, and positive for adenocarcinoma

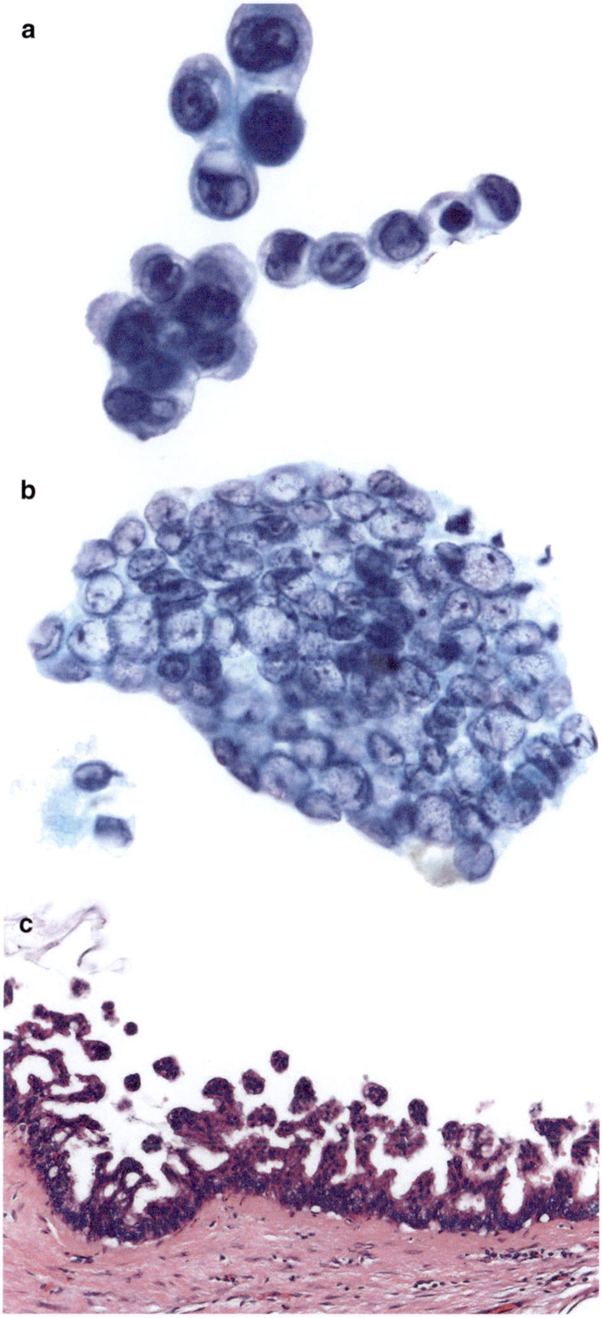
a
b
c

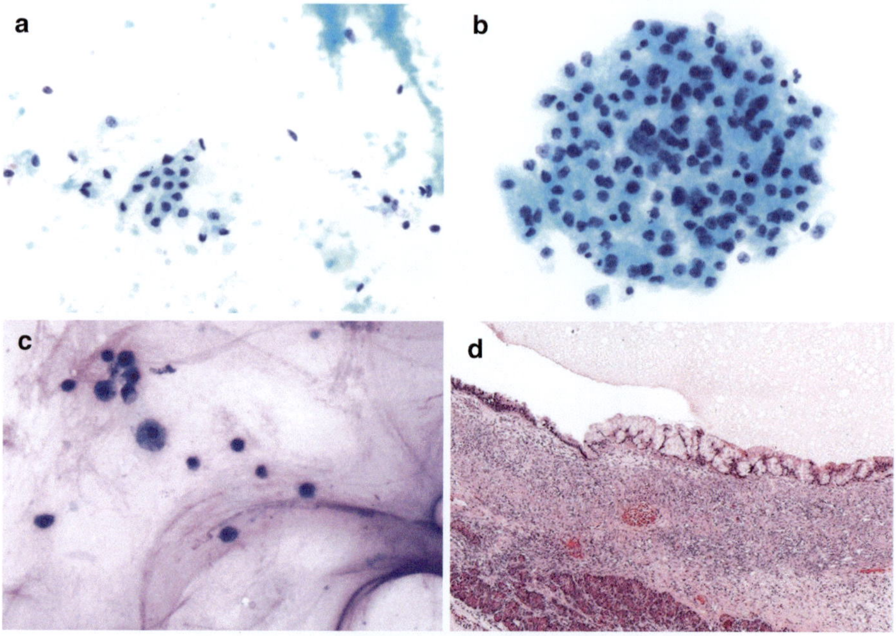

Fig. 11.9 EUS-FNA of a mucinous cystic neoplasm. (**a** and **b**) Flat honeycomb sheet of glandular cells in a background of thick mucin (**a**) and a cluster of glandular cells in a clean background (**b**). The nuclei are bland with pale chromatin and small nucleoli. Cytoplasm appears denser in the clustered cells (**a** and **b**, TP). (**c**) Isolated mucinous cells in a background of thick diffuse mucin (DQ-stained DS). (**d**) Histology showed a unilocular cyst which contained mucin that was lost on processing. The lining is of mucinous cells surrounded by ovarian-type stroma. The cyst did not communicate with the main pancreatic duct (H&E)

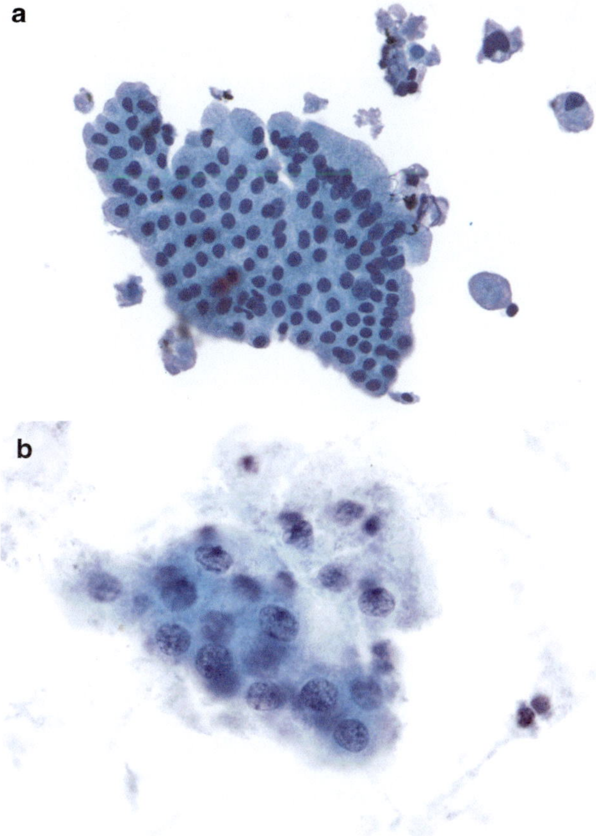

Fig. 11.10 Benign pancreatic ductal and acinar epithelium. (**a**) Benign ductal epithelium typically has a honeycomb appearance. Cells have regularly placed nuclei within the fragment, and when seen as a flat sheet, the nuclei do not overlap. Cytoplasmic mucin is not visible. Occasional variation in nuclear size can be seen, but the variation should not reach a 4:1 variation in size. (**b**) Benign acinar cells have round nuclei with regular borders, nucleolus, and granular cytoplasm (**a** and **b**, TP)

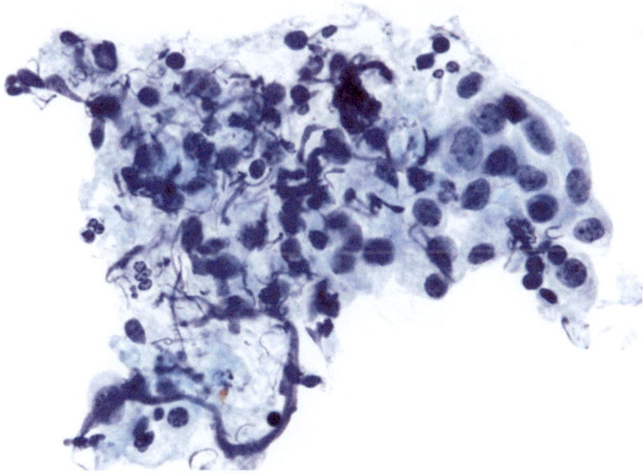

Fig. 11.11 Chronic pancreatitis: Fragment of reactive ductal cells and inflammatory cells, within fibrosis. Note reactive atypia in the ductal cells (TP)

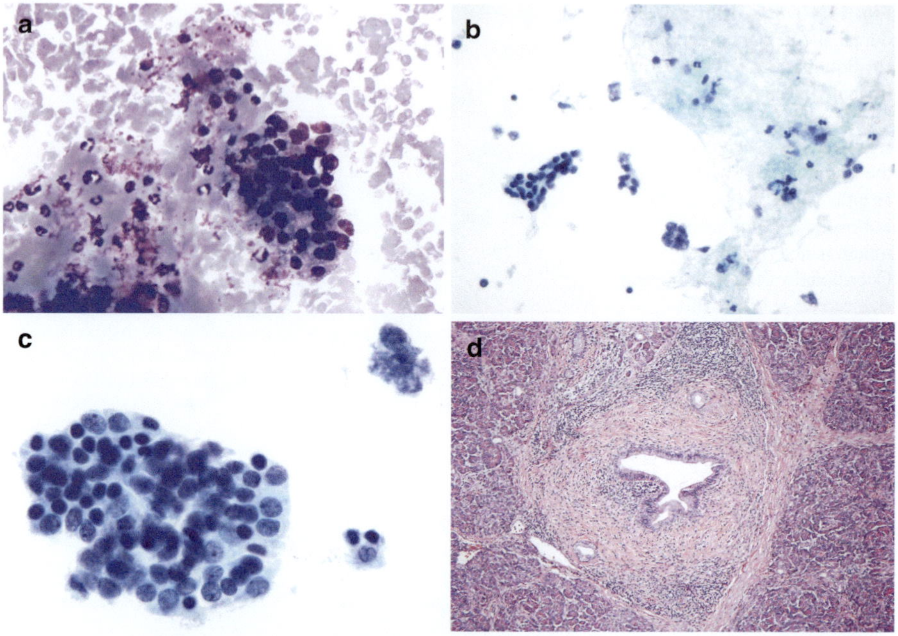

Fig. 11.12 Autoimmune pancreatitis (AIP). (**a**) Acinar cells with enlarged nuclei, granular debris inflammation, and fibrosis (DQ-stained DS). (**b**) Ductal and acinar cells in a background similar to DQ (TP). (**c**) Closer examination reveals some of the mild atypia of acinar cells that can be seen in autoimmune pancreatitis and occasionally misdiagnosed as adenocarcinoma (TP). (**d**) The resection specimen shows predominantly sclerosis with a lymphoplasmacytic infiltrate; reactive ductal cells are present as scattered glands, which may be mistaken for an infiltrative malignancy due to the distorted architecture. In IgG4-related disease, an increase in IgG4-positive plasma cells can be demonstrated with immunohistochemistry (H&E). Autoimmune pancreatitis has two subtypes; the first type is an IgG4-associated disease which affects older individuals, while the second subtype affects younger individuals and is not associated with IgG4. AIP often presents as a mass lesion, with 50 % of patients having obstructive jaundice at presentation. One series showed common features to be background lymphocytes and stromal fragments with high cellularity and embedded lymphocytes. AIP is occasionally a cause for a false diagnosis of adenocarcinoma [7]

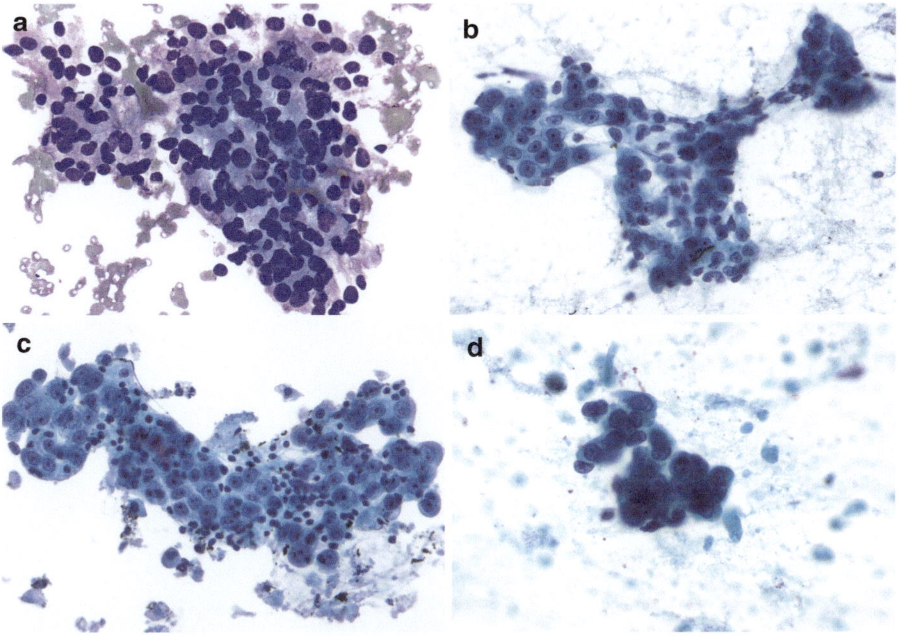

Fig. 11.13 Pancreatic adenocarcinoma. (**a**) Cells have indistinct cytoplasmic borders, but the nuclei have a disordered arrangement within the fragment. The nuclei vary greatly in size and have irregular nuclear borders and nucleoli. Cytoplasm is mucinous and nucleus-to-cytoplasmic ratio is high. A mitotic figure can be seen in the lower left corner (DQ-stained DS). (**b**) Classic cytological features of pancreatic adenocarcinoma can be seen with enlarged round to oval nuclei with thick membranes, macronucleoli, and parachromatin clearing. Cytoplasm is mucinous with distinct borders. Background shows diffuse necrosis (Pap-stained DS). (**c**) Same case in a TP preparation shows similar features except the background necrosis is clumped and clings to tumor cells (TP). (**d**) Same case in a SP preparation shows features similar to DS and TP. Some tumor cells are in focus in a given plane. Necrosis is diffuse (SP)

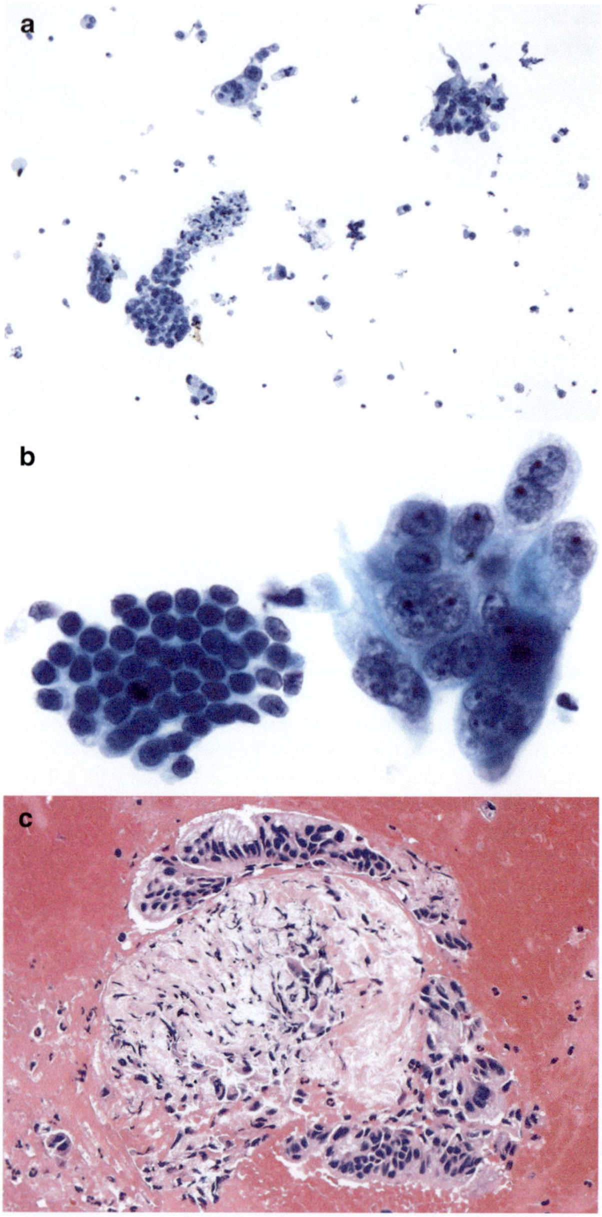

Fig. 11.14 Pancreatic adenocarcinoma. (**a**) Another case showing high cellularity and clinging diathesis (TP). (**b**) This image shows an excellent comparison of benign pancreatic ductal epithelium in a honeycomb arrangement with adenocarcinoma cells (*right side* of the field) (TP). (**c**) Cell block shows desmoplastic stroma surrounded with and some embedded tumor cells (H&E)

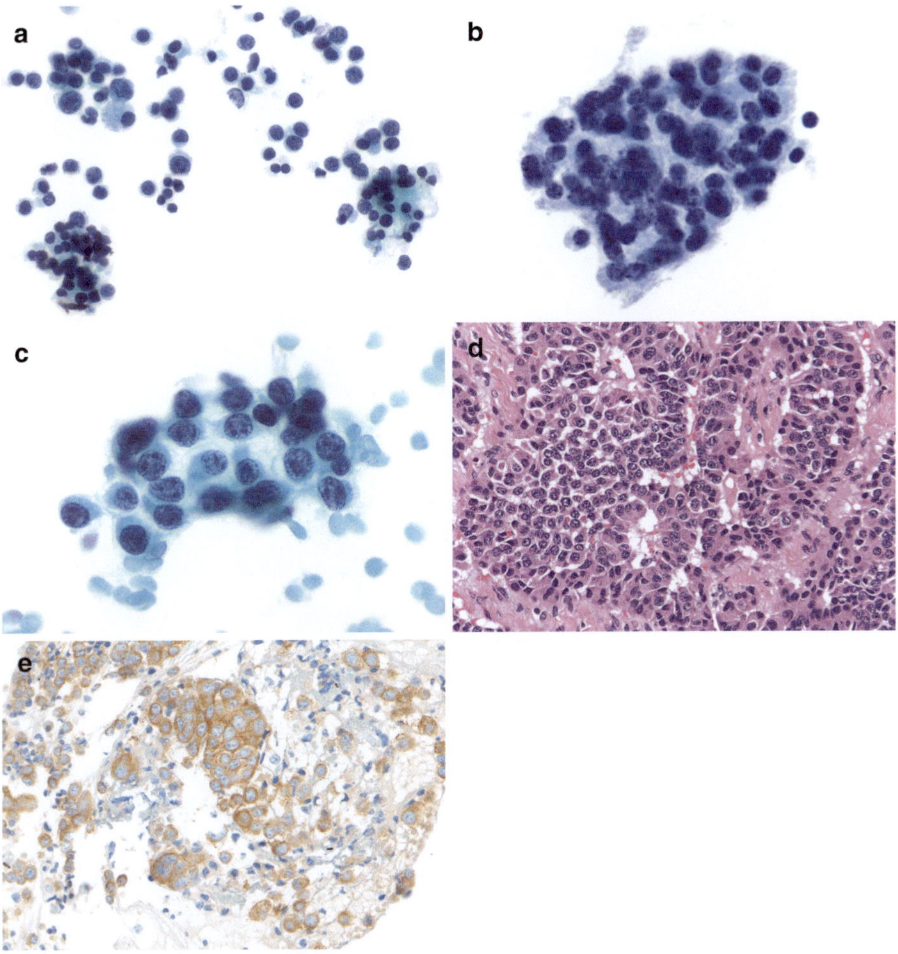

Fig. 11.15 Pancreatic neuroendocrine tumor (PanNET). (**a**) This field contains discohesive cells that form rosettes. The nuclei are eccentrically placed, variable in size, round, and regular and show a distinctive "salt-and-pepper" chromatin pattern. Cytoplasm is scant and some nuclei are devoid of cytoplasm (TP). (**b** and **c**) Nuclear features in TP are as distinct as those seen in the corresponding direct smear. Note "salt-and-pepper" chromatin and eccentric nuclei. Cytoplasm appears more granular in TP (**b**, TP; **c**, Pap-stained DS). (**d**) Resection specimen containing nests of neoplastic cells with neuroendocrine chromatin (H&E). In approximately 20 % of cases, PanNETs present as cystic lesions, in which case PanNET may not be on the initial differential diagnosis, and cyst fluid analysis is noncontributory. In such cases, FNA with cytologic examination has been shown to be an important diagnostic modality, with 71 % of cases given a specific diagnosis by cytology, compared to 47 % cases diagnosed by EUS alone [8]. (**e**) Confirmation of a neuroendocrine origin is provided by a positive immunohistochemical stain for synaptophysin. Chromogranin is another commonly used neuroendocrine marker (synaptophysin immunostain). Ki-67 showed a low proliferation index of less than 2 %

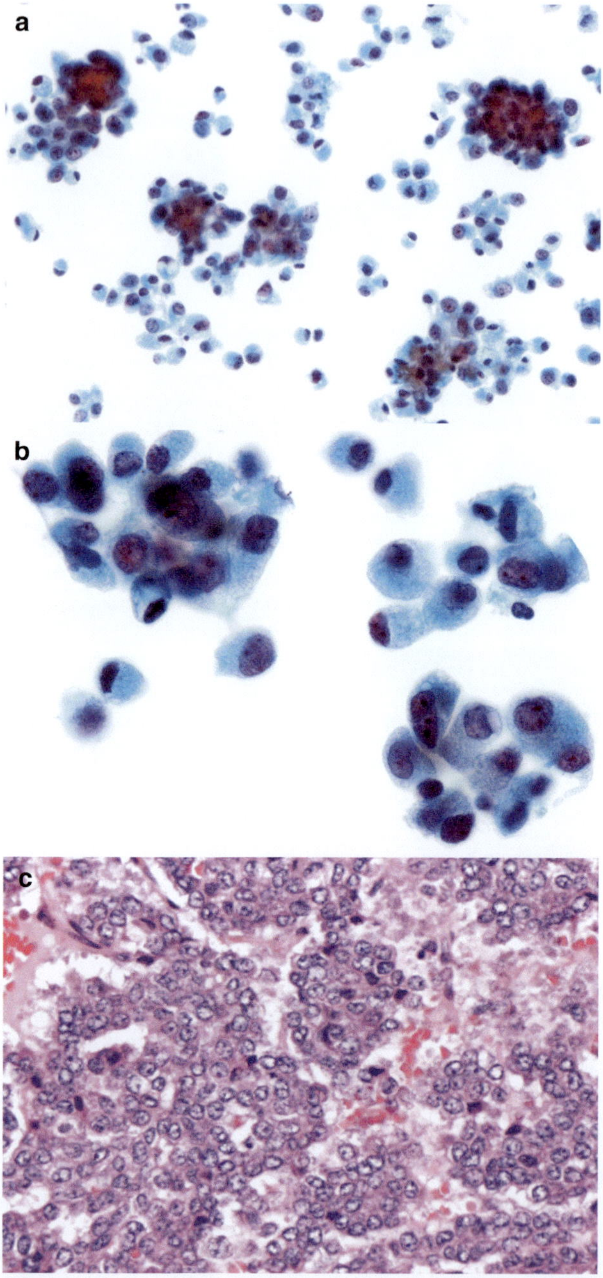

Fig. 11.16 Pancreatic acinar cell carcinoma (ACC). (**a**) At first glance, the cytomorphology of ACC appears similar to PanNET (compared to the previous figure): the cells are discohesive and form acinar structures (rosettes), and the cells have a plasmacytoid appearance. Specimen cellularity is suggestive of a neoplastic process (TP). (**b**) The cytoplasm is abundant and granular. Nuclei are round and irregular and contain coarse chromatin, cherry-red macronucleoli, and parachromatin clearing (TP). (**c**) On resection, acinar cell carcinoma forms islands of tumor cells with distinct acinar morphology. Nuclear morphology is as described for TP (H&E). Recently, BCL10 has been identified as an excellent marker for a pancreatic acinar origin, with one study showing 82 % of pancreatic ACC as positive for BCL10 by IHC, and no staining in other pancreatic subtypes except for adenosquamous carcinoma (50 % stained positive)

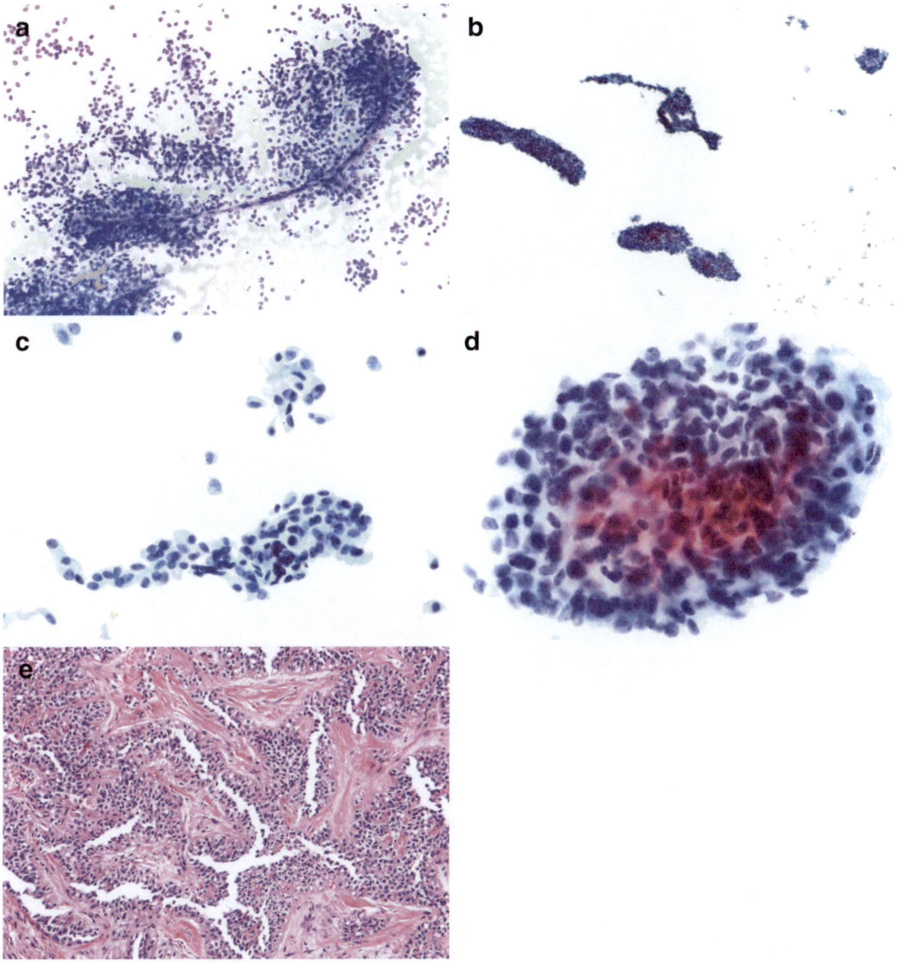

Fig. 11.17 Solid pseudopapillary tumor (SPPT). (**a**) A long vessel stretches across the field, with numerous monotonous neoplastic cells both loosely attached and scattered in the background. The cells are small and contain round and regular nuclei. The discohesive nature of the cells, as well as the plasmacytoid configuration, is suggestive of a pancreatic neuroendocrine tumor (DQ-stained DS). (**b**) TP shows a cleaner background, with fewer individual cells and predominantly irregularly shaped cellular fragments. These fragments often have unusual branching patterns, giving them unusual shapes (TP). (**c**) Neoplastic cells are monotonous with eccentrically placed uniform nuclei, small nucleoli, and moderate amount of pale-staining cytoplasm (TP). (**d**) A three-dimensional cell cluster with a central eosinophilic staining that, in TP, may represent a fibrovascular core (TP). (**e**) The resection specimen shows distinct morphology of solid pseudopapillary tumor (H&E)

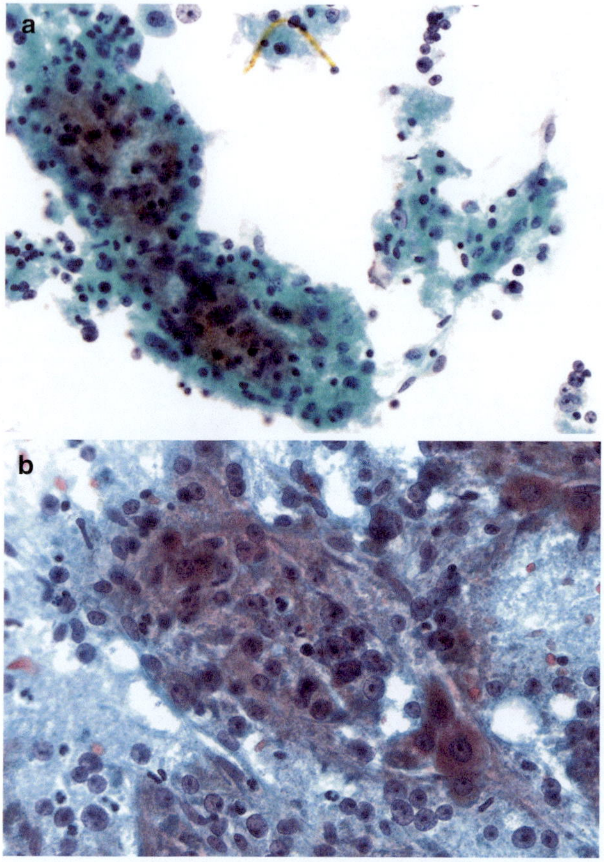

Fig. 11.18 FNA of liver with hepatocellular carcinoma (HCC). (**a**) Cluster of hepatocytes with some elongated and dark endothelial cells outlining the cluster, visible at the *left lower area*. Transgressing vessels are evident in the cells on the *right side* of the image. Nuclei are variably hyperchromatic with a prominent nucleolus; cytoplasm is abundant, dense, and granular. Background shows bare nuclei and clinging diathesis (TP). (**b**) Tumor cells have abundant cytoplasm with enlarged nuclei containing a prominent nucleolus. Endothelial cells with dark spindled nuclei wrap around the tumor cell fragment in the center of the field. Necrosis is diffuse (Pap-stained DS)

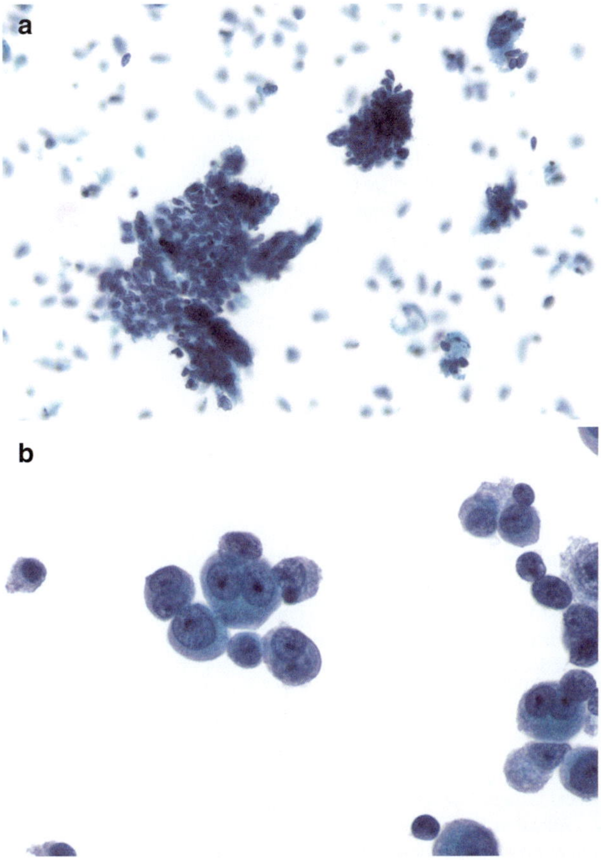

Fig. 11.19 Metastatic malignancies to liver. (**a**) Metastatic colonic adenocarcinoma shows small strips, fragments, and three-dimensional clusters of hyperchromatic and densely packed elongated tumor cells. The malignant cells have high nucleus-to-cytoplasmic ratios and hyperchromatic elongated nuclei with small nucleoli and scant cytoplasm. It is typical for colonic adenocarcinoma to have stratified nuclei in a "picket fence"-like arrangement. Here this can be seen along the edges of the fragments where elongated nuclei project out. Background necrosis, a common feature of colonic adenocarcinoma, may be greatly reduced in liquid-based preparations. Cells are in different planes of focus. The background indicates many single tumor cells (SP). (**b**) Metastatic melanoma to liver shows the classic cytomorphology with singly dispersed cells containing abundant dense to granular cytoplasm, eccentrically placed nuclei with prominent nucleoli, giving a plasmacytoid appearance. Some cells are binucleated. Intranuclear inclusions and cytoplasmic melanin pigment are not seen. Background is clean (TP). Melanoma in LBP shows similar cytomorphology as conventional smears, except the pigment and intranuclear inclusions may be reduced

Suggested Readings

1. Conrad R, Castelino-Prabhu S, Cobb C, Raza A. Role of cytopathology in the diagnosis and management of gastrointestinal tract cancers. J Gastrointest Oncol. 2012;3:285–98.
2. West RB, Corless CL, Chen X, Rubin BP, Subramanian S, Montgomery K, Zhu S, Ball CA, TO N, Patel R, Goldblum JR, Brown PO, Heinrich MC, van de Rijn M. The novel marker, DOG1, is expressed ubiquitously in gastrointestinal stromal tumors irrespective of KIT or PDGFRA mutation status. Am J Pathol. 2004;165:107–13.
3. Gurwell-Veronezi A, Wittchow RJ, Bottles K, Cohen MB. Cytologic features of villous adenoma of the ampullary region. Diagn Cytopathol. 1996;14:145–9.
4. Belsley NA, Pitman MB, Lauwers GY, Brugge WR, Desphande V. Serous cystadenoma of the pancreas: Limitations and pitfalls of endoscopic ultrasound-guided fine-needle aspiration biopsy. Cancer Cytopathol. 2008;114:102–10.
5. de Luna R, Eloubeidi MA, Sheffield MV, Eltoum I, Jhala N, Jhala D, Chen VK, Chhieng DC. Comparison of ThinPrep and conventional preparations in pancreatic fine-needle aspiration biopsy. Diagn Cytopathol. 2004;30:71–6.
6. Sigel CS, Edelweiss M, Tong LC, Magda J, Oen H, Sigel KM, Zakowski MF. Low interobserver agreement in cytology grading of mucinous pancreatic neoplasms. Cancer Cytopathol. 2015;123:40–50.
7. Deshpande V, Mino-Kenudson M, Brugge WR, Pittman MB, Fernandez-del Castillo C, Warshaw AL, Lauwers GY. Endoscopic ultrasound guided fine needle aspiration biopsy of autoimmune pancreatitis: diagnostic criteria and pitfalls. Am J Surg Pathol. 2005;29:1464–71.
8. Morales-Oyarvide V, Yoon WJ, Ingkakaui T, Forcione DG, Casey BW, Brugge WR, Fernandez-del Castillo C, Pitman MB. Cystic pancreatic neuroendocrine tumors: The value of cytology in preoperative diagnosis. Cancer Cytopathol. 2014;122:435–44.
9. Koybasioglu F, Onal B, Simsek GG, Yilmazer D, Han U. Comparison of ThinPrep and conventional smears in head and neck fine needle aspiration cytology. Turkish J Pathology. 2008;24: 159–65.
10. Hoda R. Non-gynecologic cytology on liquid-based preparations: a morphologic review of facts and artifacts. Diagn Cytopathol. 2007;35:621–34.
11. Siddiqui MT, Gokaslan ST, Saboorian MH, Ashfaq R. Split sample comparison of ThinPrep and conventional smears in endoscopic retrograde cholangiopancreatography-guided pancreatic fine-needle aspirations. Diagn Cytopathol. 2005;32:70–5.
12. Hosoda W, Sasaki E, Murakami Y, Yamao K, Shimizu Y, Yatabe Y. BCL10 as a useful marker for pancreatic acinar cell carcinoma, especially using endoscopic ultrasound cytology specimens. Pathol Int. 2013;63:176–82.
13. Jain D. Diagnosis of hepatocellular carcinoma: fine needle aspiration cytology or needle core biopsy. J Clin Gastroenterol. 2002;35(5 Suppl 2):S101–8.
14. Niimi K, Goto O, Kawakubo K, Nakai Y, Minatsuki C, Asada-Hirayama I, Mochizuki S, Ono S, Kodashima S, Yamamichi N, Isayama H, Fujishiro M, Koike K. Endoscopic ultrasound-guided fine-needle aspiration skill acquisition of gastrointestinal submucosal tumor by trainee endoscopists: a pilot study. Endosc Ultrasound. 2016;5:157–64.
15. Kang CU, Cho DG, Cho KD, Jo MS. Thoracoscopic stapled resection of multiple esophageal duplication cysts with different pathological findings. Eur J Cardiothorac Surg. 2008;34:216–8.

Fine Needle Aspiration of Breast 12

Introduction

FNA is used to obtain diagnostic material from solid and cystic breast lesions. In recent years, needle core biopsies have mostly replaced FNA for diagnosing most solid breast lesions – at least in the United States. However, FNA is still used to sample cystic breast lesions and other lesions that are difficult to access through needle core biopsy, such as in the retroareolar location. Additionally, some clinicians and pathologists continue to use FNA to sample breast lesions in the outpatient setting. Thus, virtually any breast entity can be encountered in a breast FNA.

The "triple test" is utilized to help diagnose breast lesions on FNA. The "triple test" means that the cytological findings on FNA are placed in the context of clinical and radiological findings. Relying on the results of the FNA alone can be treacherous.

FNA sampling of high-grade ductal carcinoma *in situ* (DCIS) often results in pleomorphic malignant cells that cannot be distinguished from invasive ductal carcinoma (IDC). Low-grade DCIS and even IDC contain relatively bland monotonous cells that may not reveal their malignant nature on FNA. Typically, if the clinician is suspicious of malignancy, needle core biopsies should be taken.

Upon FNA, cystic breast lesions should be drained of all fluid. If a residual solid mural component is detected, it should be sampled via FNA and needle core biopsy. Follow-up should ensure that the cystic lesion does not recur. If the cyst recurs, FNA or excisional biopsy may be required.

FNA of breast can be used to make conventional smear for rapid on-site evaluation, in which case the needle rinses can be processed as LBP. If rapid on-site evaluation is not performed, dedicated passes can be placed entirely into LBP preservative and submitted for processing.

Nipple discharge can be collected and submitted for processing. Ductal lavage can be performed to sample cells for cytological examination – if a lesion is visualized inside the duct upon galactography. In these cases, the fluid submitted to the laboratory can be processed using standard concentration techniques, including LBP.

R.S. Hoda et al., *Diagnostic Liquid-Based Cytology*,
DOI 10.1007/978-3-662-53905-7_12, © Springer-Verlag GmbH Germany 2017

Multiple studies have demonstrated the utility of LBP for breast FNA. Aside from some cytomorphological differences between LBP and conventional smears, LBP has been shown to be a reliable technique with a diagnostic accuracy equivalent to conventional smears [1, 2]. One study has shown that SP LBP provides better cellularity for cystic lesions than conventional smears [3].

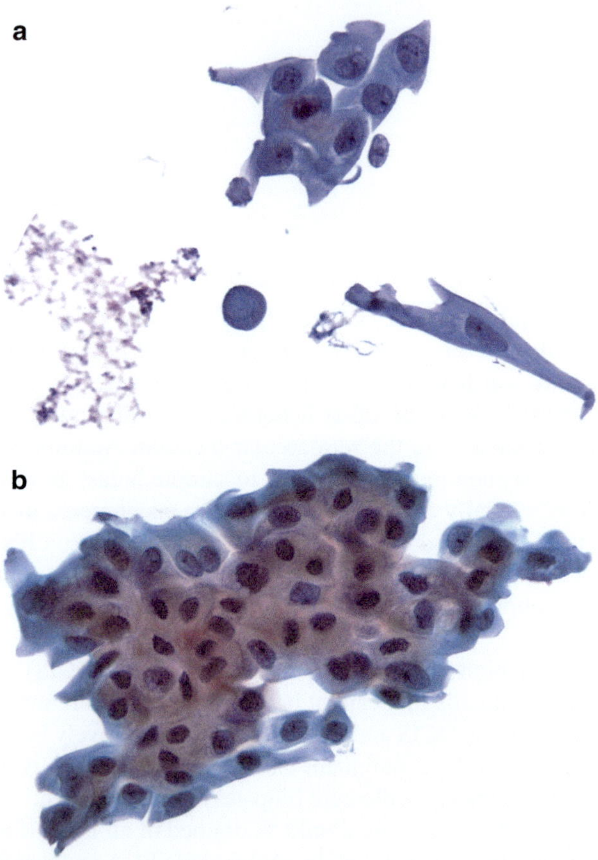

Fig. 12.1 Fibrocystic change with apocrine metaplasia. (**a**) Benign changes sampled on FNA are often typified by apocrine metaplasia. Here the epithelial cells contain abundant, amphophilic cytoplasm with round nuclei and bland chromatin. There are no nuclear irregularities, and the cells have an orderly arrangement (TP). (**b**) This larger fragment of epithelial cells demonstrates typical apocrine change; there is a minimal anisonucleosis, but otherwise the nuclei are benign-appearing, and the cells maintain an orderly arrangement (TP). Liquid-based preparation performs comparably to conventional smears for breast FNA specimen preparation and provides better cellularity for cystic lesions [3]

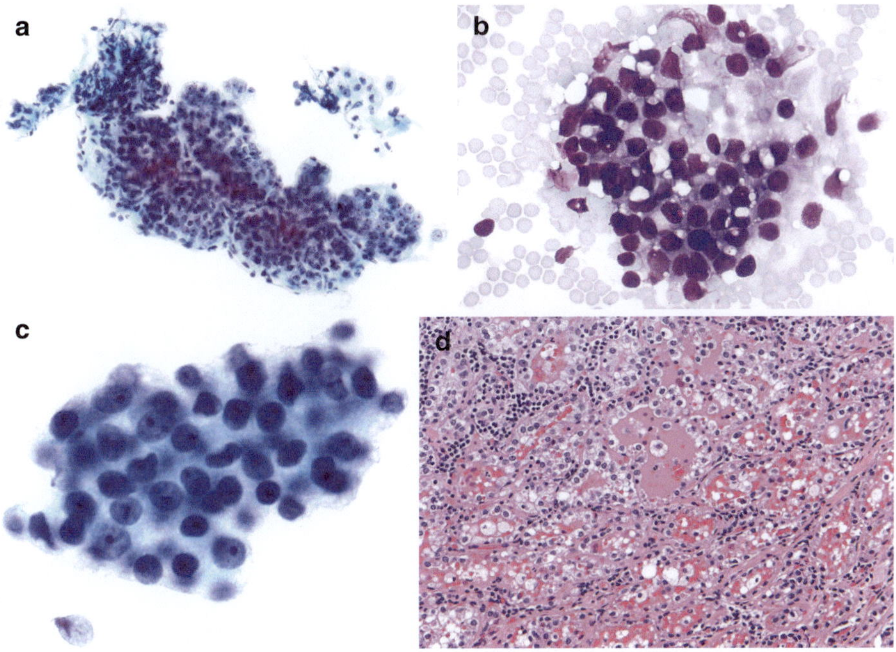

Fig. 12.2 Lactational adenoma. (**a**) Lactational adenoma typically presents as a well-circumscribed mass during or immediately after pregnancy. FNA specimens, as seen here, are typically hypercellular. The cells have foamy cytoplasm, uniform nuclei, and prominent nucleoli. Foamy material may also be seen in the background on a conventional smear, but a clean background is more commonly seen on LBP (TP). Of note, prominent nucleoli can be seen in both malignant and benign lesions [2]. (**b**) The background in this conventional smear is more bloody than foamy, but cytoplasmic vacuolization is prominent (DQ-stained conventional smear). (**c**) The cells here are monotonous, with prominent nucleoli; cytoplasm is foamy and delicate (TP). (**d**) The corresponding excisional biopsy shows hypersecretory change in mammary glands with abundant luminal secretions. The lesional cells are similar to those on cytology. (**d**, H&E)

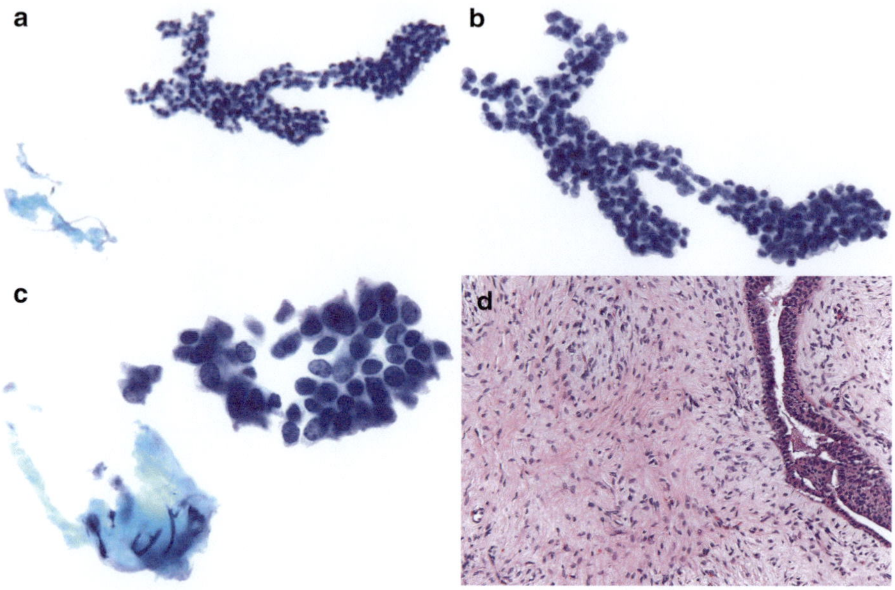

Fig. 12.3 Fibroadenoma. (**a–c**) Fibroadenoma is typified by staghorn arrangements of epithelial cells intermixed with spindled cells. The nuclei may be overlapping and can have atypia that can be mistaken for malignancy. Note the clean background. The diagnosis of fibroadenoma may be more difficult on liquid-based preparations, with studies showing a low diagnostic rate compared to conventional smears and an increased number of false-positive results [1]. The number of myoepithelial cells and stromal fragments may be reduced in liquid-based preparations (**a**,TP; **b**,TP; **c**,TP). (**d**) Fibroadenoma with elongated epithelial lined cleft amid stromal proliferation (H&E)

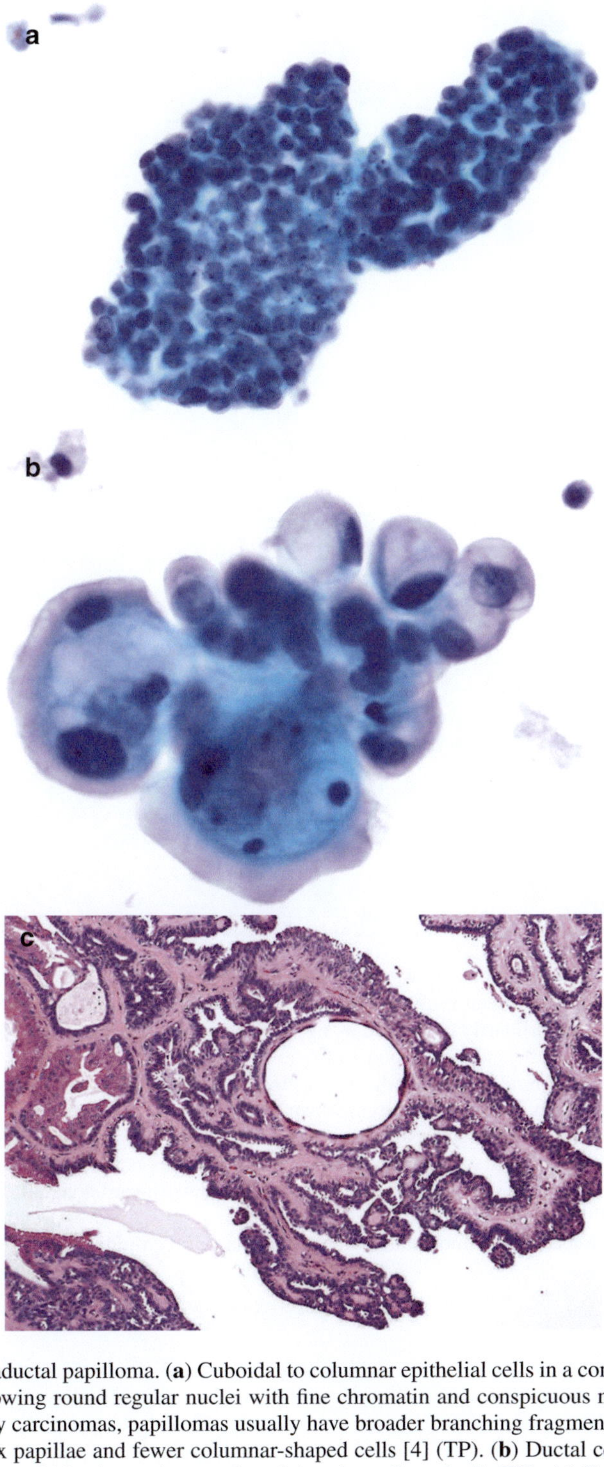

Fig. 12.4 Intraductal papilloma. (**a**) Cuboidal to columnar epithelial cells in a complex papillary-like cluster, showing round regular nuclei with fine chromatin and conspicuous nucleoli. In contrast to papillary carcinomas, papillomas usually have broader branching fragments as opposed to slender complex papillae and fewer columnar-shaped cells [4] (TP). (**b**) Ductal cells with vacuolated cytoplasm and bland round to oval nuclei and small nucleoli (TP). (**c**) Excisional biopsy specimen shows papilloma with multiple papillae containing vascularized fibroconnective tissue. Apocrine metaplasia can be seen (H&E)

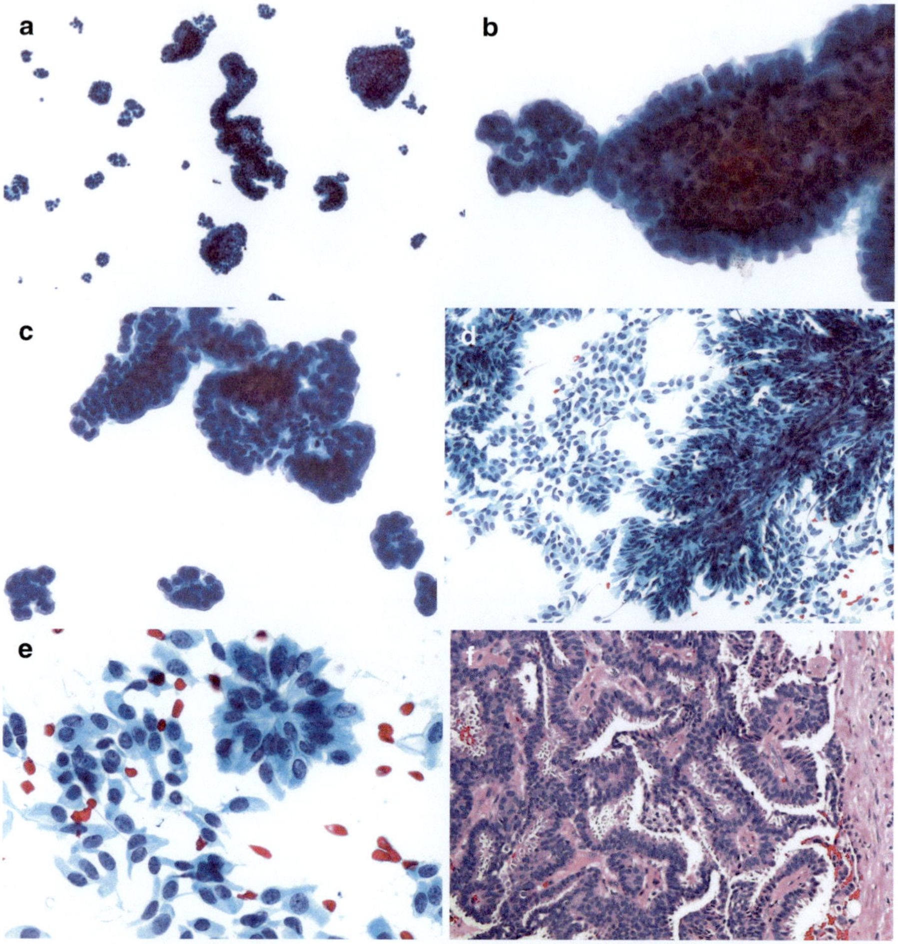

Fig. 12.5 Papillary carcinoma. (**a**) Multiple cellular fragments, some with thin branching structures, can be seen amid a clean background. Smaller clusters can also be seen (TP). (**b**) The cells around the edges are primarily columnar, as opposed to more epithelioid ones seen in (**a**). The nuclei are hyperchromatic and overlapping (TP). (**c**) The edges of these fragments are a mixture of columnar and epithelioid cells, both demonstrating hyperchromasia and anisonucleosis. Differentiating between benign and malignant papillary lesions on FNA can be challenging (TP). (**d**) Conventional smear shows a hypercellular specimen with columnar cells, many of which are loosely attached to thin papillary fibrovascular cores with complex branching (Pap-stained CS). (**e**) The monotony of cells and bland nuclear features may not suggest carcinoma; however, the specimen is hypercellular, and myoepithelial cells are absent (Pap-stained conventional smear). (**f**) Excisional biopsy specimen demonstrate complex papillae with columnar epithelium (H&E)

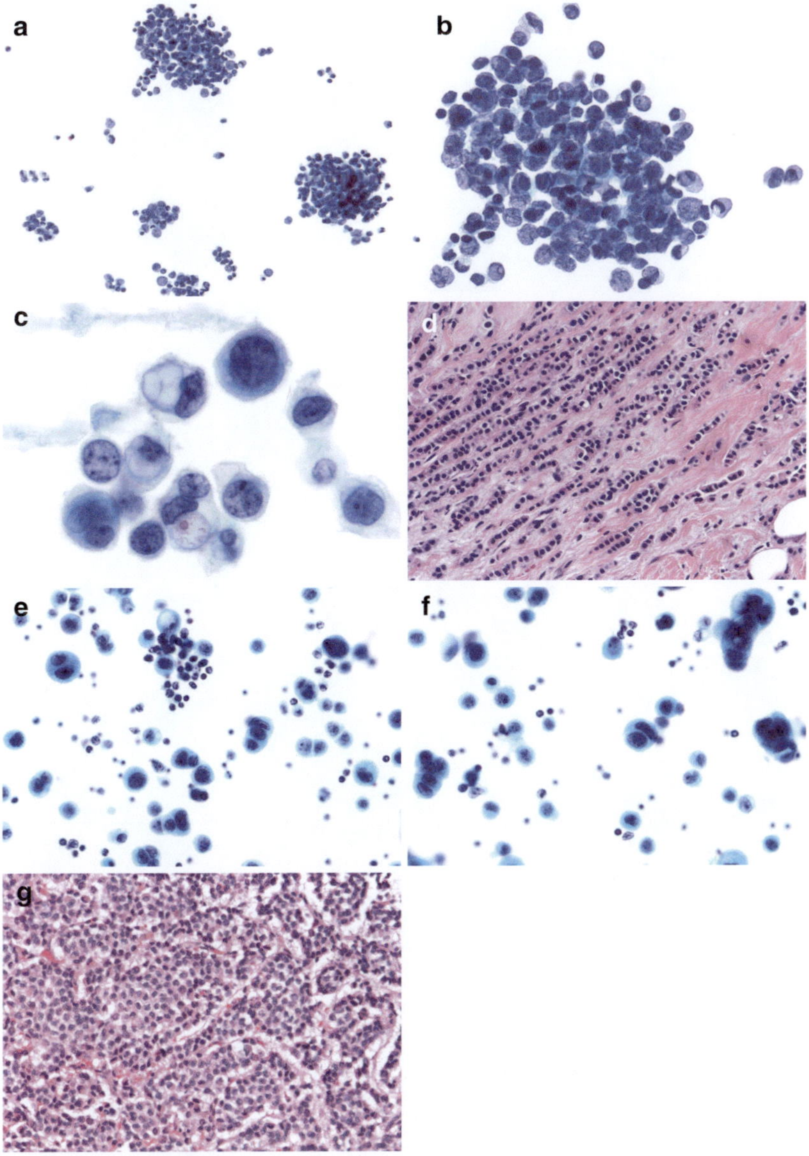

Fig. 12.6 Invasive lobular carcinoma. (**a** and **b**) Small carcinoma cells with little cytoplasm can be seen in loosely associated clusters or singly in an otherwise clean background. The nuclear borders are irregular, with moderate variation in shapes and sizes (**a**,TP; **b**,TP). (**c**) While vacuoles are not always present, some cells have large vacuoles that give the cells a "signet ring" appearance (the nucleus is compressed to the periphery by the vacuole). In this field, one cell contains condensed mucin which has stained pink on Pap stain (TP). (**d**) Invasive carcinoma cells infiltrating in a single-file manner is characteristic of invasive lobular carcinoma. (**e** and **f**) Predominantly single tumor cells with inflammatory cells in the background. Nuclei seem to "mold" where tumor cells have clustered. The background is out of focus due to the three-dimensional quality of this SP. On SP, neoplastic cells are loosely cohesive and enlarged when compared to adjacent inflammatory cells. Some cells have prominent nucleoli, and most have an eccentrically placed nucleus. The cytoplasm is foamy, with some cells showing mucin vacuoles. In some instances, the nucleus is compressed creating a "signet ring"-like appearance (SP). (**g**) excisional biopsy of case seen in (**e** and **f**) shows invasive lobular carcinoma

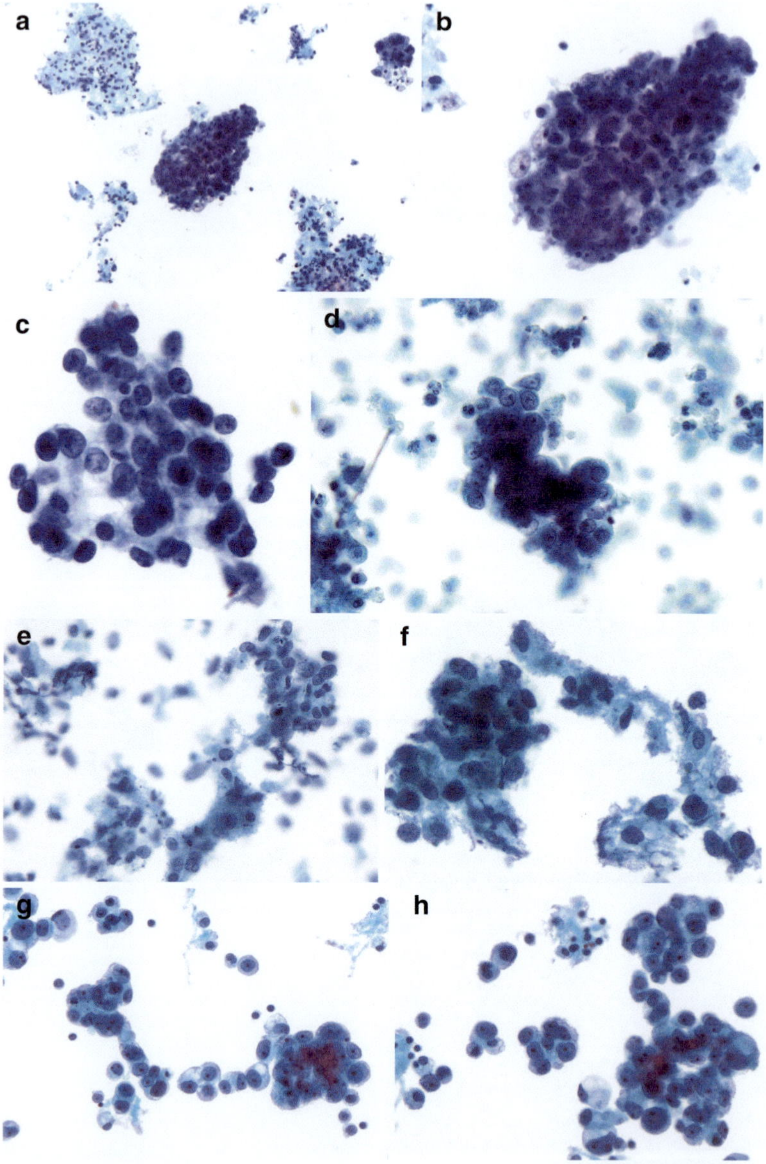

Fig. 12.7 Invasive ductal carcinoma. (**a** and **b**) Ductal carcinoma of breast shows features of adenocarcinoma seen elsewhere: clusters of three-dimensional cells with nuclear pleomorphism, hyperchromasia, and irregular nuclear borders. In this case, the cells have prominent nucleoli. Necrotic fragments are present (**a**, TP; **b**, TP). (**c**) In this image, the diathesis is not present, but the malignant nature of the cells is evident by cellular pleomorphism and nucleolar prominence. Cytoplasm is indistinct (TP). (**d**) Same case seen in (**c**), processed as SP. Notice similar cytology (SP). (**e** and **f**) Ductal carcinoma of breast processed as TP and SP. Loosely cohesive malignant cells are present amid necrosis which is more diffuse in SP and clings to tumor cells in TP (**e**, SP, **f**, TP). (**g** and **h**) High-grade ductal carcinoma of breast showing loosely cohesive glandular structures and single malignant cells. The cytology is crisp with enlarged cells, vesicular to hyperchromatic nuclei, cherry-red macronucleoli with parachromatin clearing, dense to vacuolated well-defined cytoplasm, and high nuclear-to-cytoplasmic ratio (TP). It is difficult to distinguish infiltrating ductal carcinoma from ductal carcinoma in situ on cytology alone. (**i**) Poorly differentiated invasive ductal carcinoma with high-grade nuclei and no gland formation

Fig. 12.7 (continued)

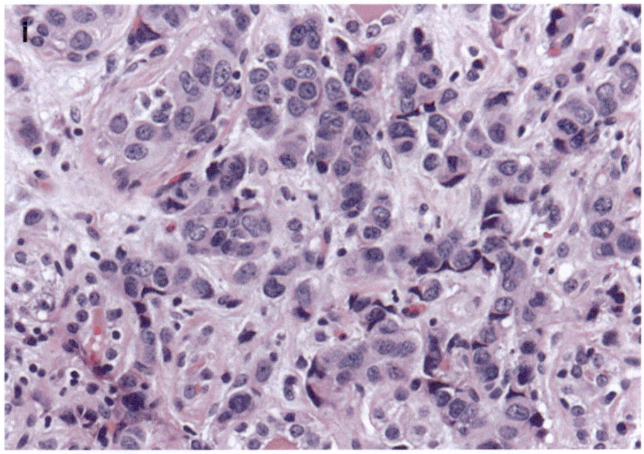

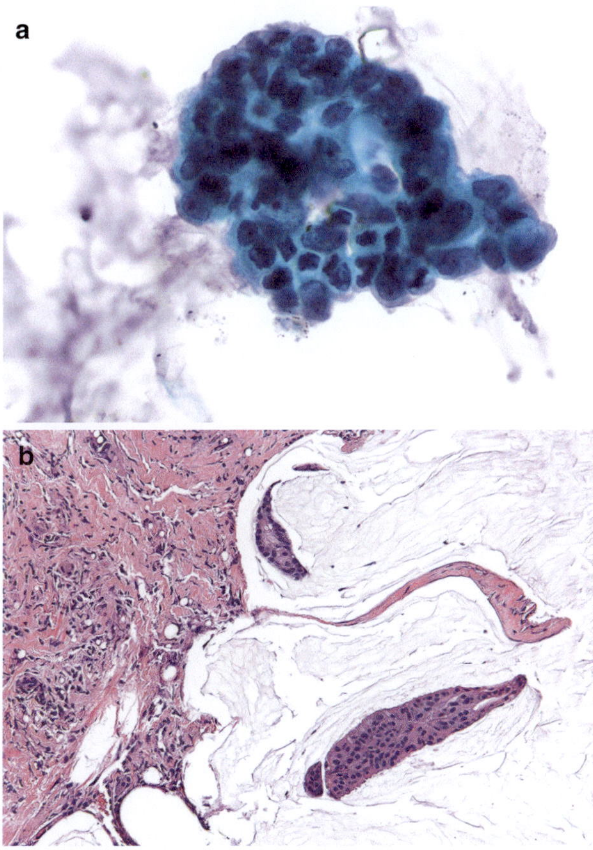

Fig. 12.8 Invasive ductal carcinoma, mucinous type. (**a**) The malignant cells have striking pleomorphism, highly irregular nuclear borders, and prominent nucleoli. Cleared-out mucin-containing area can be seen, indicating the glandular nature of the cells. Thick mucin, staining pink, is attached to the malignant cells and remains visible on LBP (TP). (**b**) The carcinoma cells are present in clusters that float amid abundant extracellular mucin

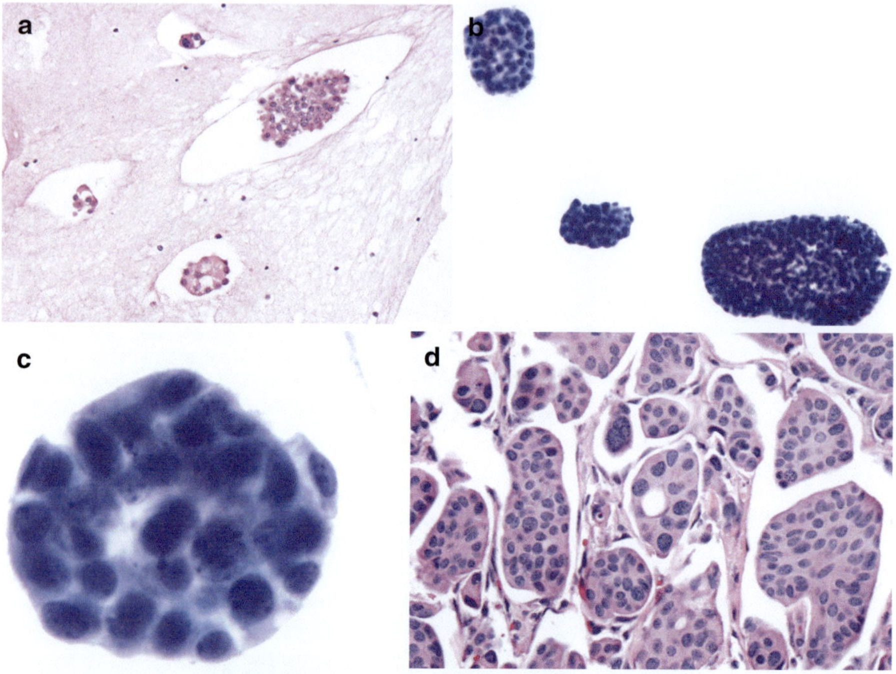

Fig. 12.9 Invasive ductal carcinoma, micropapillary type. (**a**) Cell block section shows carcinoma cells amid lacunar spaces (H&E). (**b**) The tumor cells form tightly packed balls in which the cytoplasm is indistinct and the nuclei are enlarged, and overlapping, in a clean background. The fragments appear spherical (TP). (**c**) Minimal cytoplasm can be appreciated in the tumor cells. The chromatin is coarse and there is variation in nuclear size. The borders of this fragment are extremely smooth (TP). (**d**) The neoplastic fragments appear similar on excisional biopsy with small papillary fragments forming islands of cells (H&E)

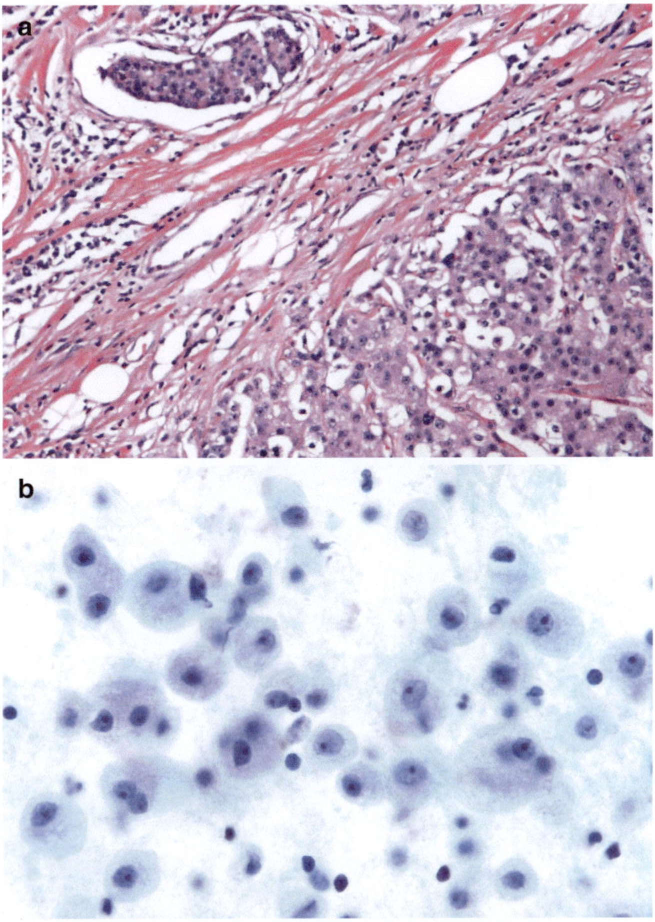

Fig. 12.10 Invasive ductal carcinoma, apocrine type. (**a**) on excisional biopsy, the granular eosinophilic cytoplasm can be appreciated. Note lymphovascular invasion by tumor cells (H&E). (**b**) loosely cohesive tumor cells can be seen with a large amount of granular cytoplasm. While the abundant cytoplasm may give a bland appearance, the enlarged nucleoli seen in these cells indicate malignancy (TP)

a

b

c

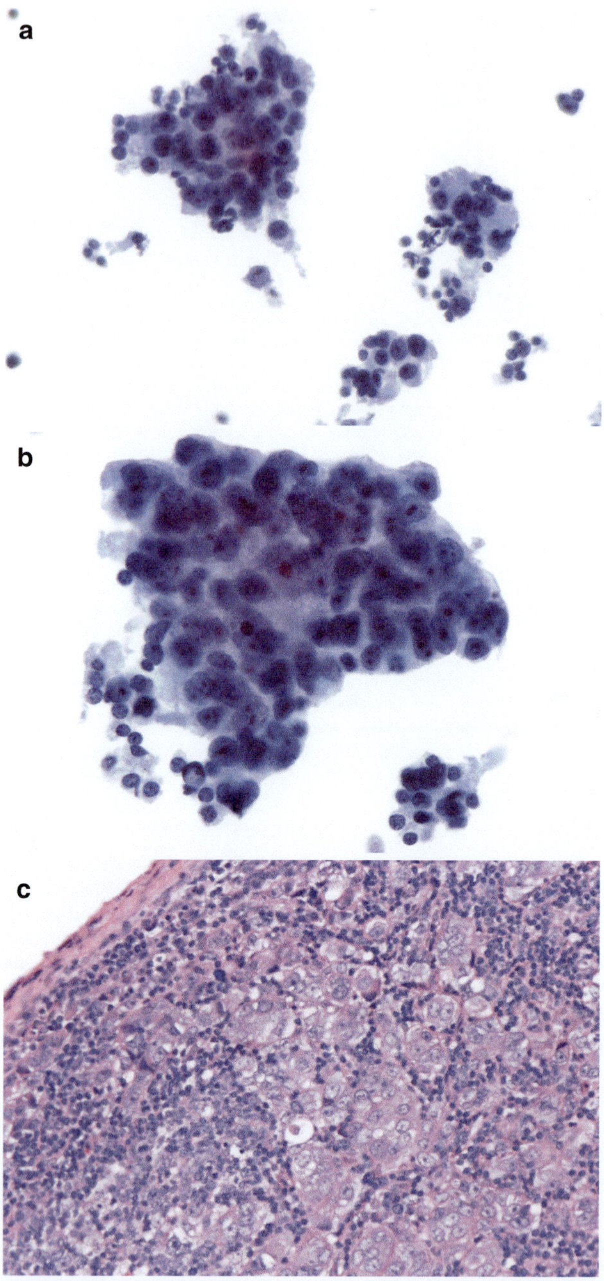

Fig. 12.11 Medullary carcinoma of breast. (**a**) The tumor cells have round and regular nuclear borders but demonstrate marked anisonucleosis as well as prominent nucleoli. The cells are both loosely clustered together and present as single cells. The cytoplasm is granular, and the eccentric nuclei have a plasmacytoid appearance. Note presence of scattered mature lymphocytes (TP). (**b**) The nuclei are enlarged and overlapping, and variation in nuclear size as well as marked nuclear border irregularities can be appreciated. Rare lymphocytes can be seen (TP). (**c**) enlarged syncytial cells can be seen admixed with mature lymphocytes, giving a lymphoepithelial appearance. Compared to LBP seen in (**a** and **b**), the lymphoid component is more prominent (H&E)

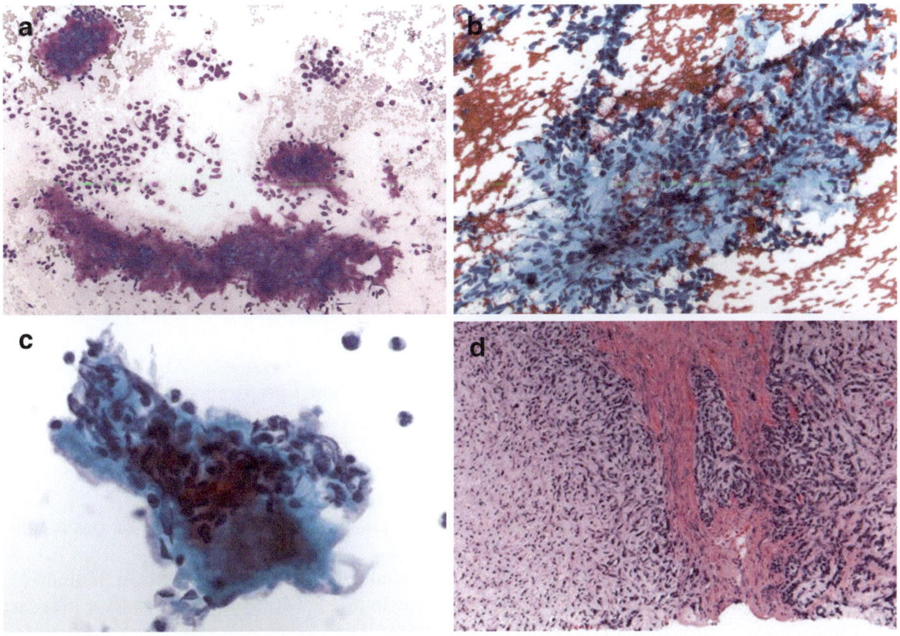

Fig. 12.12 Invasive ductal carcinoma, metaplastic type. (**a**) Ductal carcinoma of breast can occasionally be metaplastic and contain sarcomatoid elements. In this case, the carcinoma has developed an area resembling chondrosarcoma. DQ stain highlights the metachromatic "hot pink" stroma attached to pleomorphic spindle cells (DQ CS). (**b**) The stroma appears *light blue*. Spindle cells have enlarged, overlapping nuclei and are disorganized. (Pap-stained CS). (**c**) LBP has clean background but is otherwise similar to Pap-stained smear; pleomorphic spindle cells are embedded in a *blue*, thick matrix material which represents the chondroid matrix of tumor. (TP). (**d**) Corresponding excisional biopsy shows numerous pleomorphic spindle cells present within chondroid matrix (H&E)

Suggested Readings

1. Gerhard R, Schmitt FC. Liquid-based cytology in fine-needle aspiration of breast lesions: a review. Acta Cytol. 2014;58:533–42.
2. Ryu HS, Park IA, Park SY, Jung YY, Park SH, Shin HC. A pilot study evaluating liquid-based fine needle aspiration cytology of breast lesions: a cytomorphological comparison of SurePath® liquid-based preparations and conventional smears. Acta Cytol. 2013;57:391–9.
3. Eleuterio J, Aragao A, Cavalcante DIM. Adequacy of fine-needle aspiration cytology for breast lesions: the SurePath ® liquid-based technique versus conventional smears. Acta Cytol. 2015;59:253–7.
4. Michael CW, Buschmann B. Can true papillary neoplasms of breast and their mimickers be accurately classified by cytology? Cancer Cytopathol. 2002;96:92–100.
5. Yamashita A, Sakuma K, Shiina Y. Standardization of fine needle aspiration cytology of the breast—comparison of Auto Cyto Fix and conventional smears. Cytopathology. 2003;14: 79–83.
6. Hoda RS. Non-gynecologic cytology on liquid-based preparations: a morphologic review of facts and artifacts. Diagn Cytopathol. 2007;35:621–34.
7. Hasteh F, Pang Y, Pu R, Michael CW. Do we need more than one ThinPrep to obtain adequate cellularity in fine needle aspiration? Diagn Cytopathol. 2007;35:740–3.

8. Ly TY, Barnes PJ, MacIntosh RF. Fine-needle aspiration cytology of mammary fibroadenoma: a comparison of ThinPrep (R) and Cytospin preparations. Diagn Cytopathol. 2011;39:181–7.
9. Kontzoglou K, Moulakakis KG, Konofaos P, Kyriazi M, Kyroudes A, Karakitsos P. The role of liquid-based cytology in the investigation of breast lesions using fine-needle aspiration: a cytohistopathological evaluation. J Surg Oncol. 2005;89:75–8.
10. Jing X, Wey E, Michael CW. Diagnostic value of fine needle aspirates processed by ThinPrep for the assessment of axillary lymph node status in patients with invasive carcinoma of the breast. Cytopathol. 2013;24:372–6.
11. Berner A, Sauer T. Fine-needle aspiration cytology of the breast. Ultrastruct Pathol. 2011;35:162–7.
12. Dey P, Luthra UK, George J, Zuhairy F, George SS, Haji BI. Comparison of ThinPrep and conventional preparations on fine needle aspiration cytology material. Acta Cytol. 2000;44:46–50.
13. Perez-Reyes N, Mulford DK, Rutkowski MA, Logan-Young W, Dawson AE: Breast fine needle aspiration. A comparison of thin-layer and conventional preparation. Am J Clin Pathol 1994;102:349–353.
14. Mygdakos N, Nikolaidou S, Tzilivaki A, Tamiolakis D. Liquid based preparation (LBP) cytology versus conventional cytology in FNA samples from breast, thyroid, salivary glands and soft tissues. Our experience in Crete. Rom J Morphol Embryol. 2009;50:245–50.
15. Nasuti JF, Tam D, Gupta PK. Diagnostic value of liquid-based (Thinprep) preparations in nongynecologic cases. Diagn Cytopathol. 2001;24:137–41.
16. Komatsu K, Nakanishi Y, Seki T, Yoshino A, Fuchinoue F, Amano S, Komatsu A, Sugitani M, Nemoto N. Application of liquid-based preparation to fine needle aspiration cytology in breast cancer. Acta Cytol. 2008;52:591–6.

Index

R.S. Hoda et al., *Diagnostic Liquid-Based Cytology*,
DOI 10.1007/978-3-662-53905-7, © Springer-Verlag GmbH Germany 2017

MIX
Papier aus verantwortungsvollen Quellen
Paper from responsible sources
FSC® C105338

If you have any concerns about our products,
you can contact us on
ProductSafety@springernature.com

In case Publisher is established outside the EU,
the EU authorized representative is:
Springer Nature Customer Service Center GmbH
Europaplatz 3, 69115 Heidelberg, Germany

Printed by Libri Plureos GmbH
in Hamburg, Germany